The Sustainable Digital City

Avi Friedman

The Sustainable Digital City

Springer

Avi Friedman
Macdonald-Harrington Building
McGill University School of Architecture
Montreal, QC, Canada

ISBN 978-3-031-25487-1 ISBN 978-3-031-25488-8 (eBook)
https://doi.org/10.1007/978-3-031-25488-8

This Springer imprint is published by the registered company Springer Nature Switzerland AG
The registered company address is: Gewerbestrasse 11, 6330 Cham, Switzerland

Preface

In only a few decades, technological innovations have transformed the way we live and interact with one another in cities. More planners, policy makers, and private sector actors now recognize that technology and digital infrastructures can change urban life and are rapidly becoming the new frontier of urban development. By 2026, the global market for smart city infrastructure and services is expected to surpass 98 billion US$. Due to the proliferation of this industry, it is forecasted that high-tech solutions will dominate the field of urban planning in the decades to come. These changes and technological abilities are greatly needed as 70 percent of the world's population is expected to live in urban settlements by 2050.

As populations grow, society will increasingly face issues of food scarcity, air quality, and depleting fresh water supplies. Also, the growing effects of climate change caused by anthropogenic greenhouse gas (GHG) emissions will exacerbate the above issues, increase the frequency and severity of catastrophic weather events, cause biodiversity loss, and raise sea levels. With 90 percent of the world's cities being classified as coastal, these effects will be severe. Beneficial, however, is the fact that densely populated urban centers act as catalysts for technological innovation and that many solutions—both high-tech or otherwise—to these global problems are now available and will be presented in this book. The most relevant technologies to this book—the Internet, smartphones, geographic positioning systems (GPS), and other Internet of Things (IoT) devices—are presented in Chap. 1.

The driving ideas on which this book is set are largely influenced by the sustainable development goals (SDG) proposed by the UN in 2015 which underline the need for inclusive, safe, resilient, and sustainable cities and communities. For clarity, sustainability is the ability of present generations to meet their needs without compromising the ability of future generations to meet their own. As one might have already concluded, current human practices are unsustainable. To change this, the guidelines of this book are underpinned by a sustainable agenda in the economic, environmental, social, cultural, and governmental realms. Chapter 2 introduces these pillars of sustainability in the digital age and discusses how they each succeed most when the other thrives. It also introduces the four principles of sustainability and the advantages of data collection as shown in the context of ecological and carbon footprints as ways to measure sustainability.

The first domain covered is that of transportation in Chap. 3. The chapter demonstrates that the integration of digital infrastructure into urban mobility networks has the potential to redefine inter-urban transportation for the foreseeable future. When traffic sensors, streetlights, and bus lines all connect via the IoT, a smart transportation network that maximizes efficiency emerges. This chapter introduces aspects related to the integration of mobility and digital means.

Chapter 4 addresses how technology has changed the life of office workers. Within the current digital work from home (WFH) economy, employees have greater control over working hours and can redefine their work–life balance. Freed from long commutes and set schedules, many people enjoy remote work while businesses are able to eliminate the costs of operating offices. This chapter discusses how these changes have transformed the nature of work and, in turn, transferred the physical spaces required to succeed.

In response to global energy consumption expected to increase by a third and the energy sector contributing more than 70 percent of the world's anthropogenic GHG emissions, efficient energy distribution and waste management are discussed in Chap. 5. The electricity grid, smart grids, and alternative energy are introduced. Techniques of waste elimination supported by digital and other means and digitally enhanced waste-collection methods used for unavoidable waste are also covered.

Chapter 6 covers knowledge transfer, social engagement, and culture in the city. With information communication technologies (ICTs), the ways in which we learn and pass knowledge has been nothing short of transformed. This chapter discusses and profiles the ways that technology has revolutionized three different dimensions of intellectual life—social media, education systems, and cultural institutions—and analyzes case studies that demonstrate how we can leverage technology to design a more sustainable world.

Well-being, public engagement, and health care are the subjects of Chap. 7. This chapter explores comprehensively how public health and well-being can be improved using technologies. Exemplified most recently during the spread of the COVID-19 pandemic, the use of contact tracing and phone signals to predict outbreaks proves to be an effective strategy for mitigating the spread of COVID-19 and demonstrates one of the many ways in which technology can create healthier cities. Other examples such as this are explored throughout this chapter.

Chapter 8 covers home automation technologies. Scaling down the integration of digital means in cities, similar practices can be applied to the home to revolutionize efficiency and function. This chapter reviews the five most common contemporary axes of home automation technology—ideas once considered to be straight out of a science fiction novel.

The uptake of technology in the field of urban planning has allowed for the most accurate visualization of urban processes and their urban relations. This in turn allows for a better understanding of designed spaces and how they can be improved. Chapter 9 defines and discusses big-data and its relevance to urban planning and introduces geographic information systems/science (GIS), three-dimensional (3D) modelling, and virtual reality (VR).

Chapter 10 covers the digital economy and subsequent placemaking strategies. With billions of online transactions daily, fueling the world economy, there are implications for all sectors, not least urban form. This chapter further discusses the digital economy, the adaptation that retailers of different sizes must make, and the harms and benefits it poses to urban identity and placemaking.

Finally, the future of sustainable digital cities (SDCs) is explored in Chap. 11. With the ability of technology to be woven further into urban fabrics, cities can be made more comfortably livable, efficient, and sustainable in all its meanings. It highlights the importance of any size change from large-scale urban systems such as energy management and virtual health care to everyday changes such as installing smart home appliances and riding a bike.

This book offers a comprehensive overview of technological applications for sustainable smart cities and the power of the human role in actively integrating them to be used for good. As the industry will only continue to grow as technology advances, planners and professionals alike will benefit from an understanding of what these applications are today and, more importantly, what they mean for tomorrow.

Montreal, QC, Canada Avi Friedman

Acknowledgments

The material included in this book is the outcome of collaboration with and contribution by numerous colleagues, assistants, and students who directly and indirectly inspired my work. My apology if I have mistakenly omitted the name of someone who contributed to this book. I will do my best to correct such omission in future editions.

This book could not have been written without contribution to the background research, compiling information, and writing by a team of highly dedicated assistants. It included my outstanding former students Brian McGinn, Nicole Stradiotto, Genessa Bates, John Vaccaro, and Simone Tharp. Their dedication, hard work, talent, and punctuality is most appreciated.

Assembling the book's material, its organization, and copyediting were coordinated by Simone Tharp. I truly appreciate Simone's hard work, dedication, insistence on clarity, accuracy, and outstanding communication skills. Thanks also to Emma Brice and Lucy Anderson for proofreading the manuscript.

Special thanks are extended to Lucy Anderson, Elisa Costa, Jeff Jerome, Zhong Cai, Diana Nigmatullina, Jing Han (Jay), David Auerdach, Isabella Rubial, Rainier Silva, JJ Zhao, Xiong Wu Fa, Amelie Lessard, Nyd Garavito-Bruhn, Na Zhang, Maria Teleman, and Jing Yan Liu who drew the illustrations. To Charles Grégoire for the cover design and other illustrations. Their dedication, talent, and insistence on achieving excellence is truly appreciated and admired.

To Michael Luby, Senior Publishing Editor at Springer, and Zoe Kennedy, Assistant Editor, many thanks for the trust and the guidance.

Finally, my heartfelt thanks and appreciation to my wife Sorel Friedman, Ph.D., and children Paloma and Ben for their love and support.

Contents

Chapter 1
The Digital Revolution and Urbanity

In recent decades, urban growth has coincided with rapid development and integration of a variety of digital technologies into many societies. The opportunities and obstacles presented by these two phenomena complement each other in many ways. Densely populated cities act as catalysts for technological developments that are used to solve the challenges cities face. The most relevant technologies to the context of this book, the Internet, the smartphone, and satellites, have subsequently led to inventions such as the Internet of things (IoT), social media, a plethora of smart devices, and global positioning systems (GPS), among other devices. The application of these technologies to urban environments has given rise to smart cities, which use technology and other means to better meet citizen needs and improve the quality of life in cities. This chapter sets the stage for the chapters that follow by introducing these technologies, the incidental smart city transformations throughout the world, and the accompanying data privacy issues.

1.1 The Evolution of Digital Technologies

Technological developments and their integration into every facet of life in recent decades have been unprecedented. Simultaneously, urban centers are growing at astonishing rates, with 70 percent of the world's population predicted to be living in urban settlements by 2050 (UN, 2021). Each of these two phenomena poses its own unique problems and advantages. Some cities, especially the highly populated ones, act as centers of collaboration and innovation, where the best minds are called to solve the world's most challenging problems. Inevitably accompanying this innovation is crowding and traffic congestion, along with waste disposal, housing, water, and air quality needs of dense populations. These and many other social, political, and technological issues of modern-day urban living contribute to the unsustainable manner in which cities have developed in the past. The use of technology, however,

A. Friedman, *The Sustainable Digital City*,
https://doi.org/10.1007/978-3-031-25488-8_1

is being implemented so that these urban issues can now be reported, addressed, and hopefully improved. With these systems, cities can hope to function sustainably in their economies, environments, and social cultures. The ability to communicate with anyone in the world at a moment's notice afforded by Internet communication technology (ICT) has also allowed for global communication. It enabled urban centers to exchange knowledge on trials and triumphs in addressing the issues arising with constantly evolving challenges.

Historically, since the partly mechanized stocking-making machine was theorized in the sixteenth century, humans have been working to make machines work for them. The first industrial revolution in the eighteenth century saw steam engines, railways, and factory systems shift people to urban living; cities have been growing since then. The second industrial revolution beginning in the late nineteenth century is characterized by the advent of electricity, cars, and mass production, such as the semiautomated automobile factory assembly line. The most recent third industrial revolution resulted in the digitization and automation of much of the economy, thanks to the Internet. The initial stages saw inventions such as programmable thermostats, automated teller machines (ATM), and pre-set coffee makers pave the way for other technologies aimed at minimizing human intervention and redistributing tasks from humans to machinery. Later periods of the third revolution era saw the invention of the mobile phone, and, even later, tasks moved online; banking, for example, is now accessible through the Internet. Those innovations commonly followed a typical pattern that saw the introduction and use of ideas and products by a very few and their progressive adoption by many in the ensuing years (Fig. 1.1).

This rapid progression has led some to argue that today society finds itself facing what has been coined the *fourth industrial revolution*. Economist Klaus Schwab has described the present and forthcoming eras as the fusion of technologies blurring the lines between the physical, digital, and biological spheres (Schwab, 2016). Inventions such as self-driving vehicles, 3D printing, artificial intelligence (AI), energy storage, the IoT, and smart technology mean human dependence on these devices is different than ever before. The unprecedented growth of the tech industry in recent decades has allowed not only for inventions of convenience and comfort but gone beyond that to create the *potential* for convenience and comfort in a

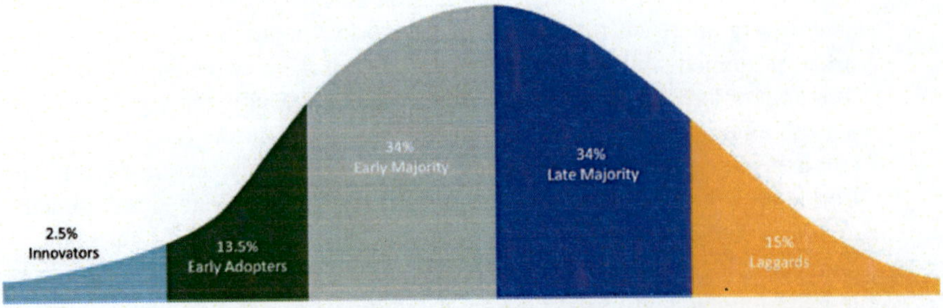

Fig. 1.1 The introduction of an innovation is commonly followed by a typical pattern that sees ideas and products used by few at first and their progressive adoption by many in the ensuing years

sustainable manner. The following subsections will highlight some of the most important technologies that have allowed for these changes and how they came to be.

1.1.1 The Internet

The Internet is a technology that allows computer networks from anywhere in the world to connect and communicate. Presently, more than 5 billion people have access to the Internet, making up more than 60 percent of the global population (Internet Live Stats, 2022). Ideas of amalgamating all human knowledge have been in the works for many centuries, and the creation of the Internet is the closest thing we have to that today. In the 1970s, the US Department of Defense developed the Advanced Research Projects Agency Network (ARPANET). This was a resource sharing network used for the purposes of communicating military resources. It was designed to perform the simple task of accepting data packets from one end and delivering them to another (Naughton, 2016). Until the 1990s, the Internet was publicly funded and used primarily for military and research purposes. The use and governance of the Internet changed in 1994 when Internet services were taken over from the US government by commercial companies known as Internet service providers (ISPs) (Naughton, 2016). This opened the use of the Internet to the public domain. At this point, anyone with a device that had the ability to connect to the Internet was allowed to do so.

The development of the World Wide Web (a so-called network of networks) in the 1990s meant that users could now publish unique content on the Internet both written and image based and proceed to interact with these media by posting comments. Other forms of communication developed such as emailing made intercommunication on the Internet more personal and direct. In 1996, Hotmail launched its email service where anyone could create their own email address for free. The notion that anyone with access to the Internet was allowed to publish content resulted in the creation of "mass media." Today, numerous industries beyond the media use the Internet for functional purposes such as transportation, healthcare, and energy management, topics that will be discussed later in this chapter.

1.1.2 The Smartphone

Pagers and pay phones quickly became ways of the past when mobile phone technology was developed. With a similar timeframe to the Internet, the first prototype of a mobile phone was released by Motorola in 1973 (Islam & Want, 2014). With two decades' worth of development later in the 1990s came a boom in the sale and use of portable mobile devices, most of which were able to make phone calls and send text messages and some of which came with a fun game or two. These phones such as *BlackBerry* had small screens and keyboards for functionality. Next, the jump from simple mobile device to smartphone came in 1999 when NTT Docomo, a Japanese mobile phone operating system, developed the i-mode system which allowed mobile devices to connect to the Internet for web browsing and email access (Islam & Want, 2014). Another milestone in the age of the smartphone was the release of the Apple iPhone in 2007. The iPhone was characterized by an iOS operating system and touch screen access, rendering the keyboard feature outdated. Not long after, Google released its operating system Android, which in 2014 held more than 80 percent of the world's market smartphone shares (Islam & Want, 2014).

Since the release of the smartphone, they have progressed invaluably with more than 80 percent of the global population now owning mobile devices (Bankmycell, n.d.) (Fig. 1.2). This has also been colored by the development of applications or "apps" that accompanied the Apple iPhone and its ensuing competitors. Most phones today come with a range of preinstalled apps; these provide basic functions such as texting, calling, music, health, organization, and more depending on the phone. Users can then download more personalized apps from their phone's respective app stores either for free or for purchase. Apple released an ad in 2009 with the slogan "There's an app for that," a seemingly self-fulfilling prophecy as now there truly seems to be an app for almost every purpose and function (Inventionland, n.d.).

Fig. 1.2 Inspecting digital appliances in a Shanghai Apple store

1.1.3 Social Media

Social media was popularized with the advent of the Internet. The universal characteristics among social media websites and applications are that users can create personalized profiles for themselves, publish and share content, and interact with other users, often in the form of comments and messages. Six-Degrees was one of the first social media sites launched in 1997, named after the theory that everyone in the world is related by 6 degrees of separation (Jones, 2015). This site and many succeeding popular social media sites, such as Myspace, Friendster, and Facebook, relied on users to make connections with their friends in which one party sends a request to another to befriend them. If the receiving party accepts, then the users are listed as friends on their respective profiles, and they are both able to see more information on the other. Many sites today however, such as Instagram, Snapchat, Twitter, YouTube, and TikTok, allow for more one-sided relationships where users can follow whoever so long as the account is listed as "public." These sites have incredible communication power with billions of users around the globe. It is common for those with large followings on social media sites, who gain celebrity status either on or off the Internet to partake in forms of "online activism" by promoting lifestyles, products, or issues they deem important. For example, actors Leonardo DiCaprio and Jane Fonda have dedicated their Internet personas to climate change activism. This can be an effective way to bring awareness of certain issues to the masses. This subject will be further explored in Chap. 6.

1.1.4 The IoT

A term used often in this book and field of study is the Internet of things, better known as the IoT. It is made up of all devices connected to the Internet, commonly known as smart devices. Self-monitoring analysis and reporting technology (smart) devices are capable of receiving data often using sensors (Techopedia, n.d.). This data is transmitted in real time through the Internet to apps or websites to be used by humans or to other devices and used autonomously. Some examples beyond laptops, smartphones, tablets, and smart watches are appliances such as lights, fridges, thermostats, security cameras, and sensors. There can also be numerous devices connected to the IoT within a single system. Jet planes, for example, contain numerous sensors and data collecting bodies that transmit information in real time to the pilot or via satellite to airlines (Aeris, 2017).

What these "things" do can vary. Fitbits, for example, track physical activity and health information from a watch worn on the user's wrist. It then transmits data from the watch over the Internet to an app either on a phone or computer for an easier visual display and record keeping. Other devices such as the Nest Thermostat have an environmentally sustainable goal of limiting energy use in the house (Google Store, n.d.). Users can synchronize their schedules to their thermostat via an app or a computer, and the thermostat then adjusts the temperature according to the schedule of the user. For example, if the house is empty from 8 am to 5 pm every day, the thermostat will turn the temperature down between those hours and ensure the house is at a comfortable temperature for the user's return. Conversely, in an office setting, an IoT thermostat could ensure temperatures are lowered overnight when employees are at home. Technologies such as this ensure the least amount of energy required is being used to heat and cool houses.

1.1.5 Satellite Technologies

A satellite is a body that orbits the Earth, moon, or other planets which can be artificial: sent into space by humans or a naturally occurring celestial body: for example, the moon is a satellite that orbits the Earth, and the Earth is a satellite that orbits the sun (NASA, 2014). Satellite technologies were developed earlier than the Internet and smartphone but came into commercial use in a similar period in the 1990s. The 1950s saw the occurrence of the space race, a competition between the United States and Russia to launch the first satellite to orbit the Earth. In 1957, Russia was triumphant in sending Sputnik 1 into space. Since then, satellite launches have become a more common occurrence with approximately 4550 documented satellites orbiting the Earth today (UCSUSA, 2022). Satellites can perform a multitude of functions, much of which come in the form of transmitting information from space about the Earth or its atmosphere back to Earth.

In 1993, satellites allowed for the advent of the GPS, a constellation of more than 30 satellites that orbit the Earth twice a day (MapToasterTopo, n.d.). They can determine the elevation, latitude, and longitude of any device that is in view of at least four satellites (i.e., most areas on Earth are in view of six satellites at any given time) (MapToasterTopo, n.d.). GPS is used for search and rescue missions, navigation, tracking, mapping, and geographic information science (GIS), a field of study to be further discussed in Chap. 9.

1.1.6 Renewable Energy Sources (RESs)

The devices and instruments that are listed in this book naturally need to be powered. Due to the climate emergency, the type of energy sources used to generate power has come under scrutiny in recent years. Traditional fossil fuel energy sources such as coal, oil, and natural gasses are large producers of greenhouse gasses (GHGs) such as CO_2. In 2015, Canada emitted 722 megatons of carbon dioxide, of which 10.9 percent of these emissions came from electricity generation alone (Canada Energy Regulator, 2020). Energy demand is also growing with an expectation that global energy consumption will increase by a third between 2017 and 2035 (Bollen, 2017). The feasibility of reducing these carbon outputs by introducing

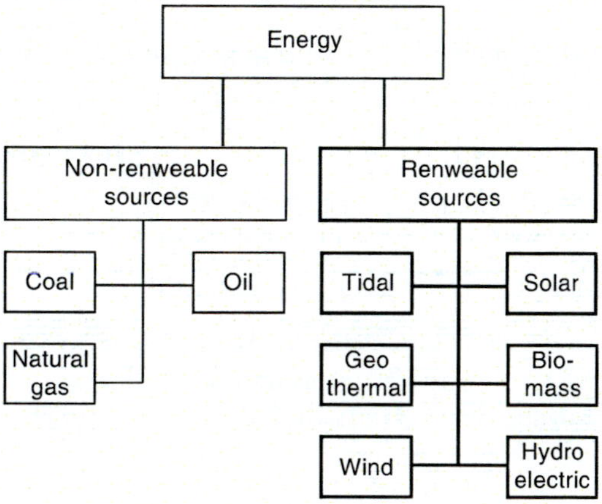

Fig. 1.3 Energy sources by type

RESs on large scales has historically been a difficult task for numerous technological, political, and financial barriers (Fig. 1.3). Time, research, and funding in the past few decades however have made the use of RESs a practical and sustainable solution to the increasing energy demand for fuel, heating, transportation, and industrial uses. More on the topic of RESs can be found in Chap. 5.

1.2 The Digital World and Daily Life

The support offered by technology has changed the way many people live their day-to-day lives. Numerous industries are also either reliant on or a benefactor of the functions of the Internet, smartphones, and satellite technology. These devices act to make life safe, efficient, and convenient. They have allowed for the best ideas from around the world to be shared, discussed, and critiqued to result in efficient and productive uses of these technologies in urban settings. This section will highlight some of the ways technology is used daily and how they can work to promote sustainable lifestyles in all sectors.

1.2.1 Interpersonal Communication

Personal relationships can be both upheld and forged by the Internet. As people immigrate and migrate around the world in an increasingly global economy, the Internet allows for the maintenance of relationships between faraway family and friends. Communication apps found on most devices today allow users not only to send text messages and make phone calls but can also make video calls allowing both users to see and hear whoever is on the other side of the call. This feature can also be beneficial to those with limited mobility making it difficult to visit friends and family. More in the realm of creating new relationships, the Internet allows for people to connect with peers that have similar niche interests who might be geographically distant. Groups using the same social media sites, chat rooms, interactive games, and more can act as accessible meeting grounds for people to interact that are not home or the workplace, otherwise known as "third spaces" or in this case cyber third spaces.

1.2.2 Work, Education, Healthcare, and Transport

In the realms of work, education, healthcare, and transportation, which will be discussed in detail in the coming chapters, the application of modern technologies has been vast. It is becoming more common in workplaces and education centers to provide employees and students, respectively, with devices that can connect them to the Internet. This has popularized online asynchronous work and learning allowing individuals to work at their own pace and gives access to online assistance such as organization tools and digital library catalogs, to name a few. In the healthcare sector, Internet-connected devices have allowed for remote activity, specifically telehealth, the benefits of which will be expanded on throughout this book. There have simultaneously been notable advances in other medical technologies such as centralizing patient care, monitors and sensors, applications of AI, vaccinations, and genome sequencing, among others (Mayo Clinic, 2020).

Lastly, technology can work to reduce GHG emissions from the transportation sector which is a major contributor to global GHG emissions. In the United States, for example, transportation was responsible for 29 percent of the country's total GHG emissions in 2019 (EPA, n.d.). There is progress being made in making transportation more sustainable, comfortable, and cost-effective. Vehicle energy sources are gradually shifting away from fossil fuels toward renewable sources such as electricity, biofuels, and hydrogen gas. There are car-share and ride-share programs that get cars off the road, lowering congestion, idling time, and thus GHG emissions. Finally, integration of GPS- and IoT-connected devices such as traffic lights is used to make travel more efficient, prioritizing buses and bikers over private vehicles on the road, making them the more convenient option for travel. Digital means and applications that support active, safe, and sustainable mobility will be discussed in Chap. 3.

1.2.3 Smart Cities

Though definitions for the term "smart city" vary, they are all underlined by the use of technology to better a city and optimize its systems (Fig. 1.4). This can be introducing technology with goals to increase economic activity, make cities more livable, initiate sustainable development, and more. Ahvenniemi et al. (2017) argue

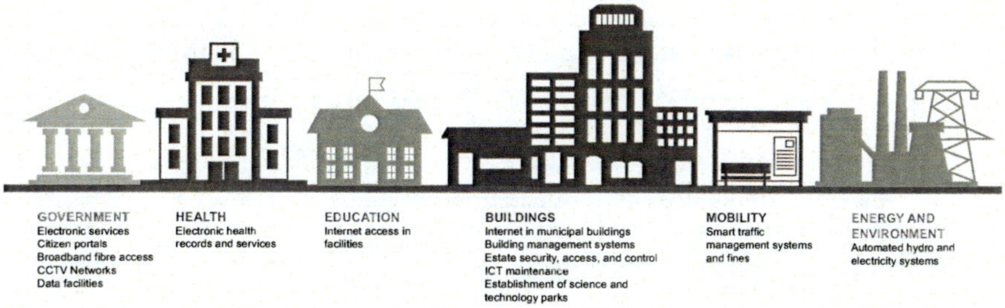

Fig. 1.4 Key features of smart cities where digital technologies are embedded in everyday life

that there are two relevant mainstreams in the smart city discussion, the first being the strictly ICT (information communication technology) and technology-oriented approach and the second being the people-oriented approach, a bettering of services, inclusion, and equity. The United Nations Economic Commission for Europe (UNECE) frames a "smart sustainable city" as a place that uses ICTs and other means to improve quality of life, efficiency of urban operation and services, and competitiveness while ensuring that it meets the needs of present and future generations with respect to economic, social, environmental, and cultural aspects (UNECE, n.d.). This definition which will be used for the purpose of this book puts an emphasis on the use of technology to increase sustainability in all the economic, social, environmental, and cultural realms. Smart cities are also responsible for bringing technology to the forefront of everyday life. Technology's role in the background of city services is far from new; sewage systems, elevators, and transportation systems as examples have been in use for decades. Some cities also have aspects of smart city technology, and many place varying emphasis on the distinct factors of sustainability. The smart city, however, requires citizens to be tuned into the digital world. Therefore, a holistic approach is needed to ensure that no organization or person is left behind during the transition into "urban smartness." Smart cities are support by the collection, analysis, and communication of data which is completed via smart sensors and other IoT-connected devices that gather real-time data (TWI, n.d.). As mentioned above, this data is then transmitted through the Internet to other IoT devices to be processed automatically and/or analyzed.

The city state of Singapore has integrated many smart city processes to make it a globally leading smart city. Singapore is consistently ranked as the top smart city for digital inclusiveness and innovative technology in global indices (Smart Nation Singapore, n.d.-b). Like many urban centers, Singapore is expecting a rapid increase of urban population over the coming decades; however, being an island, the city is constrained by a geographic limit. With a population nearing six million people and a boundary of 700 square kilometers (270 square miles), Singapore does not have the luxury of sprawling to encompass nearby peri-urban spaces and farmland. Because of this restriction, the government has been keen to adopt innovative technologies afforded by the IoT to accommodate the growing population in a forward-looking and inclusive manner.

Smart Nation Singapore has outlined that a digital society, economy, and government are the pillars of its success. These are supported by the four digital enablers that all citizens must have access to: a mobile device with network connection, Internet access, a bank account with card facility, and a national digital identity (Smart Nation Singapore, n.d.-a). To ensure no one is left behind by this rapid uptake of technology and to make it as accessible as possible, there is free Internet access throughout the island, as well as subsidies for those with disabilities and personal digital service assistance for anyone struggling to adopt or access recent technologies (MCI, 2022). Dr. Bruno Lanvin, president of Swiss business school the Institute for Management Development (IMB) Smart City Observatory, which publishes an annual smart city index, credits Singapore's achievements largely to its policies surrounding these e-government services, education, and human-centric urban strategies (Low, 2021).

Given the density of the island and various other factors, individually owned cars for the entirety of the population were and are still not feasible in Singapore. Therefore, the municipality has increased its quality and quantity of public transport systems and limited accessibility of privately owned cars to city center (Fig. 1.5). Presently, more than half of daily trips are made via public transit which is made up of a heavy rail network, light rapid transit, buses, and taxis (Calder, 2016). All these

Fig. 1.5 Readers of license plates are installed in Singapore to record the entry of private vehicles into the city center for a charge of congestion fee

systems use ICT to maximize speed and efficiency. Conversely, Singapore has a twofold technique in terms of their approach to manage and minimize private vehicles on the road (Calder, 2016). First is a vehicle quota system that limits the number of vehicles that can be owned on the island. Second is the ERP (electronic road pricing) system that taxes vehicles that drive through congestion prone areas at busy times of day. This system is enforced using transponders installed in every vehicle that charges drivers based on their vehicle type, age, and location. The results of these restrictions have resulted in a car ownership rate of 11 percent, compared to 80 percent in the United States and just under 50 percent in Europe (Budget Direct, n.d.).

In terms of food security, because of its unique landform, Singapore is presently only able to self-produce around 7 percent of its population's food requirements; the rest is imported (Krishnamurthy, 2014). To bolster independence, the city has invested in various vertical farming projects and rooftop gardens. The rotating vertical farm, for example, uses 9 meter (29.5 feet) tall aluminum towers that automatically rotate tropical vegetables from the bottom of the tower to the top (about once every 24 h) (Krishnamurthy, 2014). This automatic rotation is driven by a gravity-aided water pulley system which recycles the water to irrigate the plants. Each unit uses around 60 watts of energy per day, the same amount of energy required to power a single lightbulb. Energy in Singapore is newly powered by imported liquefied natural gas (LNG), the most environmentally friendly natural gas (Calder, 2016). Gas is cooled to a temperature of −162 °C (−260 °F) bringing it to 1/600th of its original volume and therefore cheaper and more fuel efficient to transport (National Resources Canada, 2020). Because of its geographic situation, Singapore is unfortunately not a viable candidate for wind, solar, and other RESs.

1.3 Digital Pitfalls

Smart technology has the potential to make life of citizens in cities more convenient, secure, and sustainable. Yet, the downfalls of adapting new technology cannot be unwritten. One of the biggest limitations of smart technology is that it is developing faster than it can be regulated, and outdated legislation is often not effective. Smart technology and cities also require a constant reliable Internet connection leaving many people vulnerable to power surges and Internet outages. There is the risk of incompatibility between devices themselves or devices and preexisting physical infrastructure. Location services offered on many smart devices also leave people vulnerable to tracking and location-based advertising. Additionally, any IoT device is vulnerable to malware and hacking attacks (Dwivedi et al., 2020). The cost of installing technologies can still be expensive and impractical financially for many; this poses the risk of leaving behind those who can't afford to install the newest technology, not to mention the constant costs of upkeep and updating devices. It is most notable in the developing world where accessibility to equipment and service is beyond the reach of many. The most recent United Nations Digital Economy

Report explains that the present digital divide between those who have affordable and accessible Internet access and those who do not is being exacerbated by the data-related divide (UN, 2021). Without reliable Internet access, cities cannot collect appropriate data for economic and social development placing them at a clear disadvantage. In least developed countries (LDCs), only 20 percent of people use the Internet with notable user differences in rural vs. urban areas and between men and women (UN, 2021). This divide is a growing issue as the global economy digitizes and data is increasingly valued.

Finally, one of the biggest growing concerns coinciding with the surge of IoT devices is theft, unauthorized access, or improper use of consumer data (Negash & Meso, n.d.). Data collected from consumers can be broken down into four categories: (1) personal data such as social security numbers, age, IP addresses, and device IDs; (2) engagement data, how users interact with websites, apps, and more; (3) behavioral data, purchase histories, and product usage; and (4) attitudinal data, consumer satisfaction as an example (Freedman, 2021). This data can then be used for targeted advertising, to secure more data, and/or monetized and sold by third-party data brokers (Fig. 1.6). There are numerous ways to secure data used by various institutions, many based on some of the many authentication processes to ensure the user is who they say they are (Dwivedi et al., 2020).

Data privacy also presents itself as an issue regarding public–private partnerships (PPPs) used to fund many smart cities. High infrastructure costs mean only around 16 percent of smart city initiatives can be self-funded through government taxes; therefore, PPPs are required to diversify funding sources and improve efficiency (Fishman & Flynn, 2018; Lam & Yang, 2020). Moreover, most ICT breakthroughs are achieved by the private sector, and partnerships with private entities allow easier access to these technologies (Lam & Yang, 2020). Historically, PPPs are set up between public and private entities in which the private entity fronts all or most upfront costs and the public sector repays this lease over time (Pianezzi et al., 2021). These relations however often end in governments prioritizing private needs over public ones, increased political control and influence of the private sector over public entities, and security issues concerning citizen data (Pianezzi et al., 2021). These

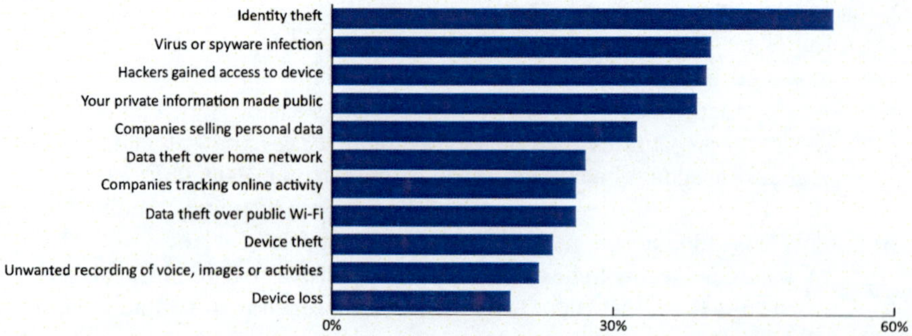

Fig. 1.6 Common consumer concerns about security/privacy issues when using digital technologies

pitfalls are very unaligned with democratic values. There are instances of successful PPPs such as that of Yokohama Smart City in Japan. Their approach was multifaceted in its success, local governments used their image to attract private interest that aligned with their local goals and values, private corporations had no expectation for returns, and the contractual agreements were more collaborative than that of a more traditional PPP (Pianezzi et al., 2021). There are also various historical and cultural relations that allowed for the smooth relations of Yokohama's PPP funding making it unique in its success.

Adopting new and changing technologies is and always has been a difficult feat. Pushback and hesitancy are to be expected. One of the biggest tools available however is education on how these technologies work, how we can use them for empowerment, what their downfalls are, and what can be done to manage these downfalls. There are many companies, for example, that believe the future of "big data" is people monetizing their own data and offering it up to corporations as opposed to it being passively accessed (Sheriff, 2019). Moving forward, this type of human-centric innovation is important. As technologies become increasingly requisite for participation in everyday life, people must be digitally literate, and technology must be as secure as possible.

Questions for a Follow-Up Discussion

1. What are the leading digital technologies that drove the fourth industrial revolution, and what was their impact on society?
2. What is a smart sustainable city, and how were its attributes manifested in the planning of Singapore?
3. What are the potential digital pitfalls that make adoption of innovation challenging?
4. Which areas in the planning and functioning of contemporary cities are going to be dominated the most by digital technologies in the coming decades?
5. Are you concerned about the role of social media in public discourse?

References

Aeris. (2017). "Connected IoT aviation and airplane monitoring". https://cdn2.hubspot.net/hubfs/246745/Aeris%20Industry%20Brief%20-%20Smart%20Airplane%20Monitoring.pdf Accessed April 6, 022

Ahvenniemi, H. et al. (2017). "What are the differences between sustainable and smart cities?". *Cities*, *60*. Part A https://doi.org/10.1016/j.cities.2016.09.009 Accessed April 6, 2022.

Bankmycell. (n.d.). "How many smartphones are in the world". https://www.bankmycell.com/blog/how-many-phones-are-in-the-world. Accessed April 6, 2022.

Bollen, M. H. (2017). "The power grid as part of a 100% renewable energy system". *Large Scale Grid Integration of Renewable Energy Sources.* Chap 1, pp 1–27. IET Digital Library. https://doi.org/10.1049/PBPO098E_ch1. Accessed April 6, 2022.

Budget Direct. (n.d.). "Car ownership Singapore 2022". *Budget Direct Insurance.* https://www.budgetdirect.com.sg/car-insurance/research/car-ownership-singapore#:~:text=Highlights,million%20vehicles%20on%20Singapore's%20roads. Accessed April 10, 2022.

Calder, K. E. (2016). "Singapore: Smart city, smart state". *Brookings Institution Press.* Chap 5. https://muse-jhu-edu.proxy3.library.mcgill.ca/chapter/1883643. Accessed April 6, 2022.

Canada Energy Regulator. (2020). "Canada's renewable power landscape 2017 – Energy market analysis". *Canada Energy Regulator.* https://www.cer-rec.gc.ca/en/data-analysis/energy-commodities/electricity/report/2017-canadian-renewable-power/canadas-renewable-power-landscape-2017-energy-market-analysis-ghg-emission.html. Accessed April 6, 2022.

Dwivedi, Y., Hughes, L., Ismagilova, E., & Rana, N. P. (2020). "Security, Privacy and Risks Within Smart Cities: Literature Review and Development of a Smart City Interaction Framework". *Information Systems Frontiers.* https://doi.org/10.1007/s10796-020-10044-1. Accessed April 6, 2022.

EPA. (n.d.). "Sources of greenhouse gas emissions." *EPA, Environmental Protection Agency.* https://www.epa.gov/ghgemissions/sources-greenhouse-gas-emissions. Accessed April 6, 2022.

Fishman, T. D., & Flynn, M. (2018). "Part two: Funding and financing smart cities series." *Using Public-Private Partnerships to Advance Smart Cities; Deloitte Center for Government Insights: London, UK*, 1–10. https://www2.deloitte.com/content/dam/Deloitte/global/Documents/Public-Sector/gx-ps-public-private-partnerships-smart-cities-funding-finance.pdf Accessed April 6, 2022.

Freedman, M. (2021) "How businesses are collecting data (and what They're doing with it)". *Business News Daily.* https://www.businessnewsdaily.com/10625-businesses-collecting-data.html. Accessed April 6, 2022.

Google Store. (n.d.). "Nest thermostat". https://store.google.com/ca/product/nest_thermostat?hl=en-GB. Accessed April 6, 2022.

Internet Live Stats. (2022). "Internet users". *Internet Live Stats.* https://www.internetlivestats.com/internet-users/. Accessed April 6, 2022.

Inventionland. (n.d.). "The history of mobile apps". https://inventionland.com/inventing/the-history-of-mobile-apps/. Accessed April 6, 2022.

Islam, N., & Want, R. (2014). Smartphones: Past, present, and future. *IEEE Pervasive Computing, 13*(4), 89–92. https://ieeexplore.ieee.org/abstract/document/6926722?casa_token=hfs_10n7E-xQAAAAA:nLpBni7XBVk9M4MEXn9aQ9u4r0wYm19QUgu0LGefqLsLpo8IAXCXWyMzwT_IT7DxsdChgqECng. Accessed April 6, 2022

Jones, M. (2015). "The complete history of social media: A timeline of the invention of online networking". *History Cooperative.* https://historycooperative.org/the-history-of-social-media/.. Accessed April 6, 2022.

Krishnamurthy, R. (2014). "Vertical farming: Singapore's solution to feed the local urban population". *Permaculture research institute.* https://www.permaculturenews.org/2014/07/25/vertical-farming-singapores-solution-feed-local-urban-population/. Accessed April 6, 2022.

Lam, P. T. I., & Yang, W. (2020). Factors influencing the consideration of public-private partnerships (PPP) for smart city projects: Evidence from Hong Kong. *Cities, 99.* https://doi.org/10.1016/j.cities.2020.102606. Accessed April 6, 2022.

Low, D. (2021). "Singapore is world's smartest city for the third year: IMD Smart City Index". *The Straits Times.* https://www.straitstimes.com/tech/tech-news/singapore-is-worlds-smartest-city-for-the-third-year-imd-smart-city-index#:~:text=SINGAPORE%20%2D%20Singapore%20is%20the%20smartest,this%20year's%20Smart%20City%20Index.&text=smart%22%20they%20are.-,A%20%22smart%20city%22%20is%20an%20urban%20setting%20that%20applies%20technology,of%20urbanisation%20for%20its%20citizens. Accessed April 6, 2022.

MapToasterTopo. (n.d.). "How GPS works". *MapToaster. https://www.maptoaster.com/maptoaster-topo-nz/articles/how-gps-works/how-gps-works.html.* Accessed April 6, 2022.

Mayo Clinic. (2020). "Telehealth: Technology meets health care". Mayoclinic.org. https://www.mayoclinic.org/healthy-lifestyle/consumer-health/in-depth/telehealth/art-20044878. Accessed April 6, 2022.

MCI. (2022). "Digital readiness". *Ministry of Communication and Information.* https://www.mci.gov.sg/en/portfolios/digital-readiness/get-digitally-ready. Accessed April 6, 2022.

NASA. (2014). "What is a Satellite". https://www.nasa.gov/audience/forstudents/5-8/features/nasa-knows/what-is-a-satellite-58.html. Accessed April 6, 2022.

National Resources Canada. (2020). "Liquified natural gas". *Government of Canada.* https://www.nrcan.gc.ca/our-natural-resources/energy-sources-distribution/clean-fossil-fuels/natural-gas/liquefied-natural-gas/5679. Accessed April 6, 2022.

Naughton, J. (2016). The evolution of the internet: From military experiment to general purpose technology. *Journal of Cyber Policy, 1*(1). https://doi.org/10.1080/23738871.2016.1157619. Accessed April 6, 2022.

Negash, S., & Meso, P. (n.d.). "Privacy concerns in the age of smart devices". Federal Trade Commission. https://www.ftc.gov/system/files/documents/public_comments/2017/11/00028-141795.pdf

Pianezzi, D., Mori, Y., & Uddin, S. (2021). Public–private partnership in a smart city: A curious case in Japan. *International Review of Administrative Sciences.* https://doi.org/10.1177/00208523211051839. Accessed April 6, 2022.

Schwab, K. (2016). "The fourth industrial revolution: What it means, how to respond". *World Economic Forum.* https://www.weforum.org/agenda/2016/01/the-fourth-industrial-revolution-what-it-means-and-how-to-respond/. Accessed April 6, 2022.

Sheriff, L. (2019). "This app enables you to make money off your own personal data". *Forbes.* https://www.forbes.com/sites/lucysherriff/2019/03/29/this-app-enables-you-to-make-money-off-your-own-personal-data/?sh=5d9f7aa429f6. Accessed April 6, 2022.

Smart Nation Singapore. (n.d.-a). "Pillars of a smart nation". https://www.smartnation.gov.sg/about-smart-nation/pillars-of-smart-nation. Accessed April 6, 2022.

Smart Nation Singapore (n.d.-b). "Achievements". https://www.smartnation.gov.sg/about-smart-nation/our-journey/achievements. Accessed April 6, 2022.

Techopedia. (n.d.) "Smart device". *Techopedia Dictionary.* https://www.techopedia.com/definition/31463/smart-device Accessed April 6, 2022.

TWI. (n.d.). "What is a smart city? - Definition and examples". *TWI Global – The Welding Institute.* https://www.twi-global.com/technical-knowledge/faqs/what-is-a-smart-city#:~:text=For%20example%2C%20smart%20parking%20can,by%20a%20smart%20city%20infrastructure. Accessed April 6, 2022.

UCSUSA. (2022). "UCS satellite database". *Union of Concerned Scientists.* https://www.ucsusa.org/resources/satellite-database. Accessed April 6, 2022.

UN. (2021). "Digital economy report 2021". *United Nations Conference on Trade and Development.* https://unctad.org/webflyer/digital-economy-report-2021 Accessed April 10, 2022.

UNECE. (n.d.). "Sustainable smart cities." UNECE. https://unece.org/housing/sustainable-smart-cities#:~:text=A%20smart%20sustainable%20city%20is,as%20well%20as%20cultural%20aspects. Accessed April 6, 2022.

Chapter 2
A Need for Sustainable Urban Environments

2.1 Global Social Phenomena

Continuous mismanagement of natural resources has stressed many natural systems to their limits. This, paired with growing populations and urbanization trends, has likewise stressed anthropogenic systems. As a result of these phenomena, cities are at a turning point soon to be home to more than 70 percent of the global population (UN, n.d.-a). Resolving these contemporary global challenges requires changing past unsustainable habits with the help of innovative technologies and mindsets. The implementation of sustainable practices, those that ensure the ability to "meet needs of the present world without compromising the ability of future generations to meet their own needs," is regarded essential (UN, n.d.-b). Thankfully, present-day digital capabilities allow for cities to sustainably adapt their soft and hard infrastructure to accommodate bigger populations and lead the way in tackling many of the global challenges to be outlined below.

2.1.1 Air Pollution and Climate Change

The overconsumption of natural resources is a contemporary pressing global issue. While there are some technological solutions to these problems, lifestyle and priority changes are needed to sustainably manage essential resources. To begin, according to the World Health Organization (WHO), 99 percent of the world's population breathes air that exceeds pollution levels with low- and middle-income countries experiencing highest exposure levels (WHO, n.d.). Urban environments pose unique risks of air pollution exposure to their populations because of "urban canyons" formed between buildings which trap and concentrate pollution from vehicles and other sources. The most promising solutions for improving air quality are the

A. Friedman, *The Sustainable Digital City*,
https://doi.org/10.1007/978-3-031-25488-8_2

cessation of burning fossil fuels and its accompanying policies, proper waste management, and the use of clean power sources (WHO, n.d.).

In terms of food security, with historical fluctuations, food insecurity of severe and moderate levels now affects 30 percent of the population, and more than 3 billion people lack access to healthy diets (FAO, 2021). Tackling food security challenges means strengthening all economies, intervening in food supply chains to lower the cost of nutritious food, ending poverty, and changing consumer behavior among other large feats. Water, on the other hand, is a finite resource and a right to life, second only to air. While it covers 71 percent of the Earth's surface, only 3 percent of all water on Earth is fresh, while the other 97 percent is salinated ocean water (Water Science School, 2019). Uneven distribution of natural freshwater and limited resources paired with increasing populations mean that water scarcity is a growing concern and has the potential to cause economic, social, political, and security conflicts on a global scale (Contento, 2018). The impacts of water scarcity due to increasing populations will be felt as early as 2025 most significantly in northern Africa, Eurasia, the Middle East, and the United States. Looking further forward, by 2050, freshwater scarcities will affect nearly 5 billion people (Campbell & Tindall, 2010). Solutions to water security include desalination of ocean water, reuse of water, decentralizing resources, and decreasing demand through water efficient systems (Global Water Security, n.d.). The move to urban environments marking postindustrial lifestyles keeps people miles (both figuratively and literally) away from the production and management of resources. In the United States, for example, the year 1900 saw 41 percent of the population providing food to the nation compared to 1.8 percent today (Contento, 2018). As evident, a reconnection to food and water sources and knowledge of air quality are key elements to sustainable urban environments. Successful solutions will use relevant technology to solve the problem directly and to encourage lifestyle and habit changes, similar to those solutions proposed to mitigate climate change (Figs. 2.1 and 2.2).

Climate change is the long-term shift in global temperatures and weather patterns caused by greenhouse gasses (GHGs) such as carbon dioxide enhancing the greenhouse effect on Earth (UN, n.d.-d). While some attribute these changes to cyclical patterns of solar radiation, humans have been the main drivers since the first industrial revolution when the burning of coal started being used as an energy source. Since then, anthropogenic emissions have been increasing so much that global warming is projected to reach 3.2 °C (5.8 °F) by the end of the twenty-first century (UN, n.d.-d). Resultant effects are melting polar ice, rising sea levels, biodiversity loss, unpredictable weather patterns, and an increase in frequency and severity of catastrophic weather events, among others (UN, n.d.-d). Climate change will also exacerbate the abovementioned complexities of food security, water security, and air quality and will disproportionately affect those living in smaller and global south nations hardest and first. Technological solutions to climate change however are present in abundance, and many must now simply be adopted.

The burning of fossil fuels as an energy source—one of the biggest contributors of GHGs to the atmosphere—can be replaced with clean energy sources such as wind, solar, and hydro- or geothermal energy which will be further discussed in

Fig. 2.1 Agricultural production in dessert conditions is made possible by using digital irrigation technologies

Chap. 5 (Fig. 2.3). Similarly, fossil fuel-powered transportation can be exchanged for public and active transit and electric/clean energy-powered vehicles (Fig. 2.4). With the research having been done and numerous solutions to mitigate the effects of climate change being presented, the need now is only to act. Some barriers to action, however, are lack of finances, political polarization of climate change, and the private sector unwilling to put climate action before profit, hence the need for the prioritization of sustainability.

As technology advances, the human needs for adequate food, water, air, and shelter remain the same and represent some of the direst challenges today. There are numerous technological solutions available such as optimizing food production and distribution, water desalination, and air quality monitors, to name a few. However, these phenomena require a more holistic approach. The following sections will present the concept of sustainability and how a well-rounded, sustainable approach to the future of urban growth is the most promising way to meet present and future population needs.

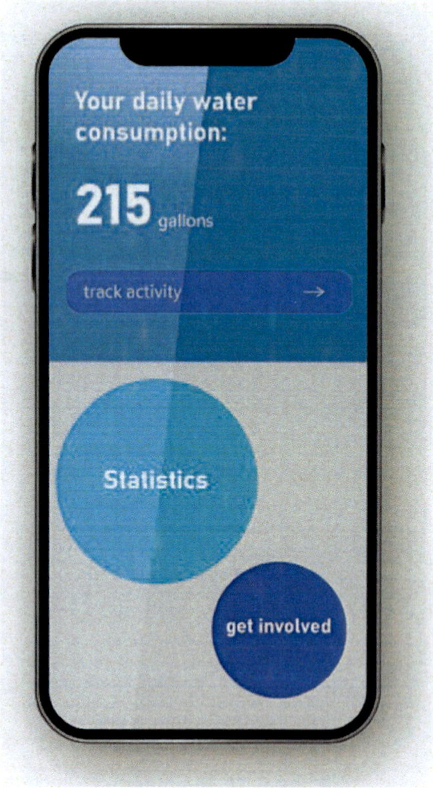

Fig. 2.2 This application helps occupants monitor their daily water consumption to encourage lifestyle and habit changes

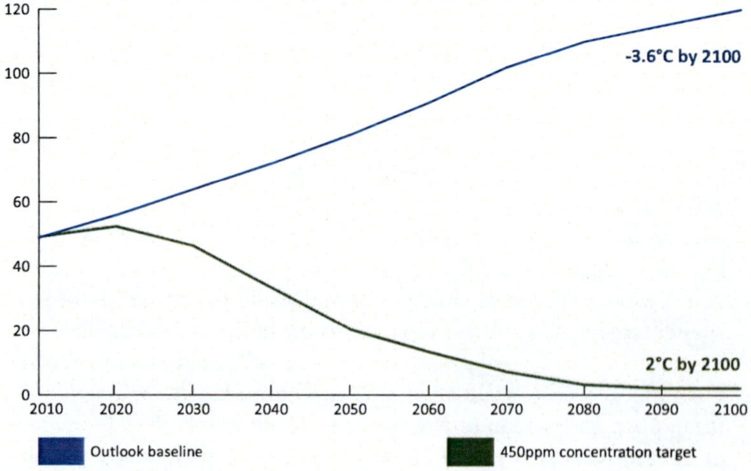

Fig. 2.3 Present projection of GHG emissions

Fig. 2.4 A charging station for electric vehicles sends a message to the owner when the process is complete

2.2 The Genesis of Contemporary Sustainable Thought

The roots of the term *sustainability* can be traced back to 1972 at a UN conference on the human environment in Stockholm, Sweden. This was the first international meeting devoted exclusively to climate change, which addressed the ongoing environmental damage and the future of humanity (Alalouch et al., 2019). It also gave notice to the gap in consumption between the global north and south and increased population growth straining natural resources including air, water, land, flora, and fauna (UN, 1972). The conference resulted in participants adopting a series of principles for sound management of the environment such as the Stockholm Declaration and Action Plan for the Human Environment. In 1987, after the 1972 UN conference failed to have a major impact on mitigating environmental degradation, the Brundtland Report was published covering yet again the threats brought to the planet by under- and overdevelopment (Keeble 1988). As noted above, it is the first documented use of the term *sustainable development*, defined as "meeting the needs of the present without compromising the ability of future generations to meet their own needs" (UN, n.d.-b). The Brundtland Report highlights the interconnectivity of the environmental, social, and political spheres, growing poverty rates, and the need for socially and environmentally sustainable economic growth specifically in the global south (Fig. 2.5). This holistic approach has underpinned definitions of sustainability and sustainable development ever since.

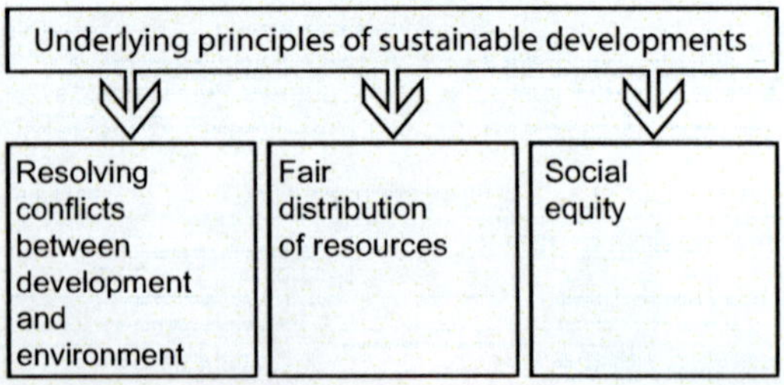

Fig. 2.5 The Brundtland Report highlights the growing poverty rates and the need for equity in realizing socially and environmentally sustainable economic growth specifically in the global south

Fig. 2.6 In 2012, the SDGs were introduced at the UN conference in Rio de Janeiro

Since the publication of these reports, there has been much progress in international agreements surrounding sustainability and climate change. In 1988, the Intergovernmental Panel on Climate Change (IPCC) was established to provide governments of all levels with credible scientific information on climate change in all its forms so that the most effective policies could be implemented (IPCC, n.d.). They provided transparent scientific papers and assessment reports reviewed by experts. In 1997, the Kyoto Protocol was established, committing nations to reduce emissions of GHGs. In 2012, the Sustainable Development Goals (SDGs) were introduced at the UN conference in Rio de Janeiro (Fig. 2.6). There are 17 SDGs, all of which recognize "that action in one area will affect outcomes in others and that development must balance social, economic and environmental sustainability" (UN, n.d.-c). These goals range from gender equality and conserving the oceans to building resilient infrastructure that promotes sustainable industrialization, all acknowledging that no one can occur without the other (UN, n.d.-c). Finally, in 2016, the present geological era starting in 1950 to the present day was coined the *Anthropocene*, a title indicating that human activity is and has been the dominant influence on climate and the environment (Pavid, n.d.). In this regard, there is no denial of the presence and impact humans have on the Earth.

Aside from international agreements, literature surrounding sustainable practices has also progressed significantly since 1972. There are currently numerous models of sustainable businesses, development, and lifestyles. Since the year 2000, 5000 books have been published with the word "sustainable" or "sustainability" in their title, compared to zero before 1976 (Caradonna, 2018). For example, Stephen Wheeler in *Planning for Sustainability* (Wheeler, 2004) further addresses the interconnectedness of economic, social, and environmental concerns. He lists goals of equity, environment, and economy as the "Three E's" to characterize an integrated approach to planning. Wheeler explains that the least developed "E," equity, is fought most by policy makers hoping to concentrate wealth and power (Wheeler, 2004). Responses to literature such as this have attempted to address the inherent relation between economic, social, and environmental success, that one without the other is not feasible. Harvard Business School, for example, encourages a practical approach with their sustainable business strategy: partnering with nonprofit organizations, educating employees, encouraging volunteerism, evaluating supply chains to source materials responsibly, reducing consumption of natural resources, and reducing carbon emissions (Stobierskie, 2021). The private sector has an undue amount of power to persuade actors big and small to think sustainably, and these suggestions, while not exhaustive, demonstrate that sustainability in all realms has value.

Sustainability in the media is dominated by coverage on climate change, and its interdisciplinarity of it is often lost. In 2006, climate change activist and former US Vice-President Al Gore released the documentary film *An Inconvenient Truth*, warning watchers of the human-made issues of global warming and climate change. Greta Thunberg, a teenage Swedish activist, has gone viral for holding weekly "school strikes for climate change," to pressure governments to reach carbon reduction goals. Other documentaries, articles, and more, reporting the risks of climate change, have saturated the media in recent years. Sustainability is a broad topic, and it must be accurately portrayed that many systems, issues, and thoughts are part of the environment and need to be sustained (Friedman, 2007). While scholars, activists, and the media may differ in approaches on how best to achieve sustainable economies, environments, and societies, there is a consensus that imminent action is required to change the present trajectory negatively affecting the Earth.

2.3 The Key Pillars of Sustainability

The Brundtland Report identified *social, economic, and environmental* sustainability as the three original pillars of sustainable development, with *culture and governance* being added soon after to ensure holistically sustainable initiatives (Fig. 2.7). The addition of culture and governance ensures that developments such as new urban buildings must be designed with local sociopolitical context in mind. This section discusses these five pillars of sustainable development that act as a guidance

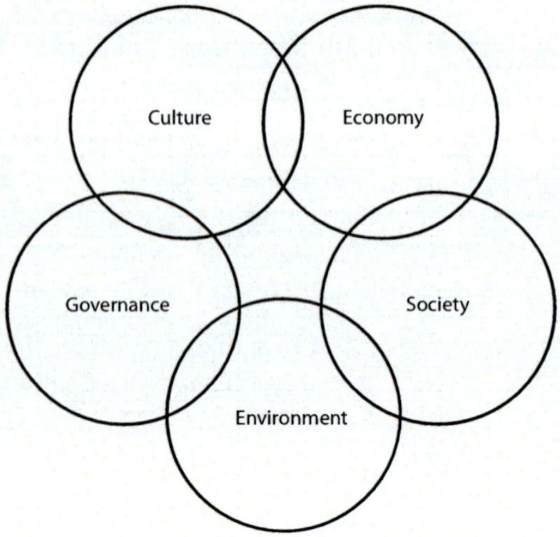

Fig. 2.7 The Brundtland Report identified social, economic, and environmental sustainability as the three original pillars of sustainable development with culture and governance being added soon after

for planners and establish the roles in which newly developing technology can play in supporting each of these pillars.

2.3.1 Social Sustainability

Social sustainability is symbiotic with economic, environmental, cultural, and governing sustainability yet often overlooked. Concerned with the development of social capital, resilient communities, human rights, public health, labor rights, and social supports, social sustainability encompasses social equity and sustainable communities (Arnett, 2017). Its complexity, however, does not lead to a single definition. Sustainable communities, outlined as a feature of social sustainability, are recognized by Arnett (2017) as the significance of social networks, community activities, sense of place, and residential stability and security in helping to sustain communities at a neighborhood level. The growing automation of tasks such as self-checkouts at grocery stores, online banking, shopping, and self-driving vehicles instilled by the top-down planning of many smart cities can leave citizens isolated and out of decision-making processes. Digital citizen participation (DCP) in which governments of smart cities actively engage in feedback and dialogue is one of the many ways technologies are used to enhance social sustainability. Alalouch et al. (2019) attempt to fill the literature gap on social sustainability within smart cities and outline a twofold connection between the digital era and social sustainability. First, it allows for better quantification of the "soft" qualities that define community satisfaction such as the sense of identity, social cohesion, happiness, or quality of life. Data collected from DCP allows for quantitative observations such as number of participants, their demographic information, and nature of comments to be translated to softer themes such as equity, inclusion, democracy, empowerment, identity, pride, and social innovation. Second, DCP allows for community engagement, which in turn leads to social sustainability and an overall sustainable model for cities.

2.3.2 *Cultural Sustainability*

Promoting cultural sustainability, such as the preservation of local traditions, arts, music, buildings, and vernacular culture, contributes to a sustainable society in both direct and indirect ways (Fig. 2.8). As explained by the Canadian Commission for UNESCO, the 2030 Agenda for Sustainable Development highlights the significance of culture as it was woven throughout the 17 SDGs that were mentioned above in areas such as climate action, poverty reduction, and food security (Canadian Commission for UNESCO, 2020). Culture contributes to a shared history and shapes values, perspectives, and identities which dictate actions. The effects of cultural artifacts and traditions can be reported, monitored, and quantified by digital means. Further, as the digital world increasingly shapes everyday life and practices, it inherently defines the present global culture.

Fig. 2.8 Cultural initiatives such as the preservation of local traditions, arts, music, and buildings contribute to social sustainability in both direct and indirect ways

2.3.3 Economic Sustainability

Economic sustainability is one of the most prioritized pillars of sustainability influenced by both the public and private sectors. The economy itself is the process by which goods and services are produced, sold, and bought. However, a sustainable economy is one that uses, safeguards, and sustains resources to create long-term sustainable values through optimal use, recovery, and recycling (Merriam-Webster, n.d.; University of Gävle, 2018). Moreover, a sustainable economy ought to avoid the transfer of costs resulting from bad present decisions to future generations. Ethical, sustainable business practices have become increasingly popular. As a result, public pressure has pushed businesses to invest in green technologies and develop more sustainable business practices (Pfeffer, 2010). Businesses that have changed to be environmentally and socially conscious benefit from increased public support and therefore increased revenue. The role of technology and the digital world in supporting sustainable and local economies like farmers' markets lies in its fast communication abilities to drastically influence the advertising world among other industries (Fig. 2.9). For example, a company that wishes to ameliorate their marketing strategy can collect relevant data on consumers based on their age, location, and income and send out targeted advertisements. This allows for more

Fig. 2.9 The role of digital means in supporting sustainable local economies like farmer markets lies largely in its local and regional reach out abilities

efficient communication and means consumers receive the most relevant information and products for themselves and are more likely to make purchases. Another example lies in the power of social media. If someone purchases a sustainable product and deems it likable, they might post something on a social media account for their followers to see. This acts as free advertising for the brand and brings in income from followers who decide to purchase from the brand.

2.3.4 Environmental Sustainability

Environmental sustainability is what comes to people's minds upon mentioning the term sustainability. Characterized by defining a boundary for us to satisfy our current needs without compromising the quality of the environment/ecosystem so that it remains equally capable of supporting future generations too, environmental sustainability aims to preserve natural resources among other goals (Bajya et al., 2019). A *cradle-to-cradle cycle* can be used to ensure the environmental sustainability of new developments and will be further discussed later in the chapter (Fig. 2.10). Today, environmentally sustainable technologies tend to focus on rectifying the consequences of past unsustainable actions by creating sustainable solutions for the present and future in all industries. Shoes and clothing made from plastics, recycled material, and even waste are some ways in which excess used material is being repurposed in a functional manner. Likewise, boots made from mycelium that can be composted upon their end of life are another way in which the fashion industry is approaching the *cradle-to-cradle* or circular economy approach. Alternatively, new digital apps and websites such as FoodRescue.ca in Canada are aiming to reduce food waste, increase food security, and mitigate climate change by connecting grocery stores, restaurants, and other food providers with nonprofit and social service organizations (Second Harvest, n.d.). Food providers send less if any food to landfills (which in turn produce GHGs), and nonprofits can get safe food for free. This also promotes social sustainability, supporting local individuals who lack food security (Sitra, 2020).

2.3.5 Inclusive Governance

Lastly, sustainable governance is key to promoting other sustainable policies. Sustainable governance aims to develop sustainable policy outcomes and imbue political decision-making with a longer-term focus (Schraad-Tischler & Seelkopf, 2015). High performing sustainable governments demonstrate sustainable policies, robust democracies (electoral processes, access to information, and civil rights), and good governance consisting of executive and media accountability and competence, evidence-based action, and adaptability, among other attributes (SGI, n.d.). The evolving needs of smart cities require the attention of an adaptable smart

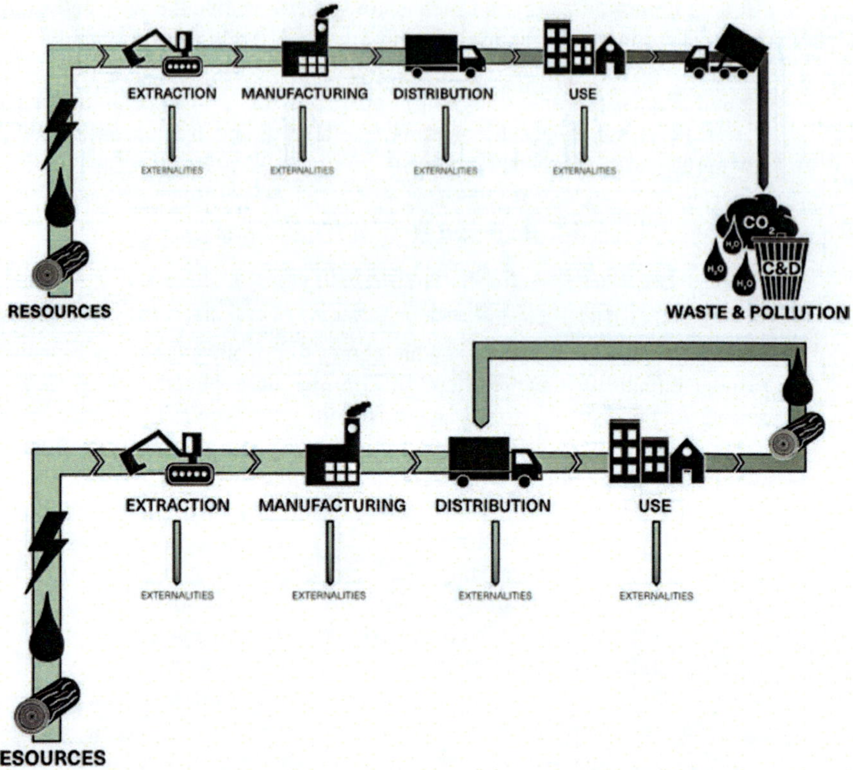

Fig. 2.10 Whereas a conventional economy supports a take, make, and dispose model (top), a cradle-to-cradle model sees resources and products stay in circulation for as long as possible (bottom)

government. For example, Singapore has made policy co-creation via technology-enabled data platforms a norm, allowing non-state actors to easily contribute to mutually beneficial policies (Woon, 2018). Open data platforms that consolidate and share publicly available data sets, apps that promote citizen feedback, and crowdsourcing portals such as hackathons and competitions allowing citizens to contribute to solving urban problems are all ways in which Singapore has, and other governments can, embrace digital platforms. The city also has digital administrative centralization for organization, communication, and record keeping of internal and external government affairs.

2.4 Principles of Sustainability in the Digital Context

Rapid urbanization in the present and coming years is forcing expansion in most urban areas. Historical practices such as urban sprawl have resulted in negative implications for all the abovementioned pillars of sustainability. Moving forward, sustainable urban development can be looked at as a process rather than a single finite project. The use of the following four axioms of sustainability with the aid of digital means can guide planners in these processes (Fig. 2.11).

2.4.1 The Path of Least Negative Impact

This first principle, the *path of least negative impact or resistance*, argues that a decision-maker or planner must act in ways that will have the least negative impact on all social, cultural, economic, environmental, and government systems. One of the most vital steps in ensuring sound decision-making throughout a project is performing an impact assessment at the front end of the planning process. This will ensure any short-term and/or long-term consequences are kept to a minimum if not reduced altogether. In the digital age, geographic information systems (GIS) prove to be useful tools as they can aid in the impact assessment and planning processes of development projects. An example would be choosing the location of a new school to ensure it is as walkable as possible. GIS tools can help planners locate residential areas, main streets, and transport opportunities for students and ensure the school is located at the best available property according to the listed criteria. Similarly, if planning for the location of a new wastewater treatment facility for an urban space, vegetation, population, street, and water catchment area data can be overlaid. The comparison of this data can help planners locate the facility as close to the city as possible for short transport times benefitting economic and environmental sectors but as far away as possible from wildlife, green space, human

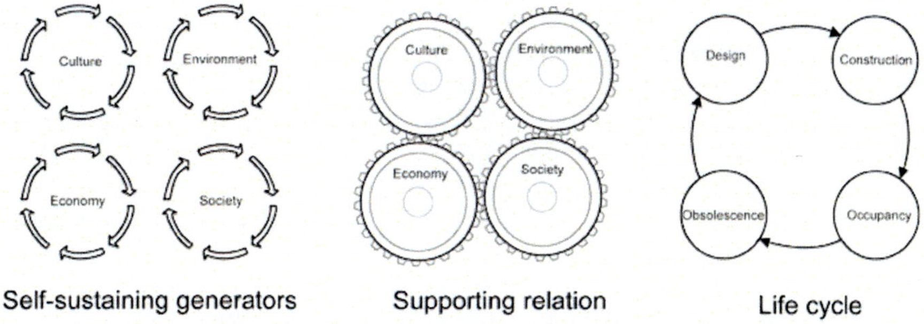

Fig. 2.11 Sustainable urban planning and development can be looked at as a process wherein key disciplinary principles are supported by digital means to guide planners in decision-making

dwellings, and water sources for social and environmental benefit again. The uses of GIS for urban planners will be further expanded in Chap. 9.

2.4.2 Self-Sustaining Processes

Instilling *self-sustaining processes* of resources and activities is the second principle of sustainability. Reliance on a single economic or energy source, for example, has the potential to strain a system. If the source fails, the system's integrity is jeopardized, and long-term planning is made difficult. Self-sustaining processes intertwine numerous sources and backup plans to ensure resilience. An example of this as mentioned in the previous chapter is the use of numerous clean energy sources for a city. A mix of solar, wind, and hydropower can all be used interchangeably, so that if the weather is not sunny or windy, hydropower can act as a substitute. Another example of self-sustaining processes is the decentralization of services. As climate change increases the severity and frequency of extreme weather events, the ability to hybridize services is becoming more important. Excess rainwater, for example, can cause flooding, especially in largely concrete environments that are not equipped with proper drainage measures. Rainwater harvesting is an easy way to capture stormwater. It can be used for washing and toilets or be cleaned with biofilters and sterilized to make it potable (University of Toronto, 2017). This also changes the water source away from a traditional centralized municipal water supply. Adding other hybrid approaches to rainwater management can come in the form of green roofs and walls and bioretention basins such as rain gardens or bioswales (University of Toronto, 2017).

2.4.3 Supporting Relations

The third principle of sustainability is the development of *supporting relationships* between the five pillars of sustainability. The goal is for the pillars to strengthen and support each other resulting in a supportive system. An improvement in one of the pillars will propel improvement in the others and likewise with a failing pillar. For example, the construction of dense low-energy-use buildings proves to be environmentally sustainable. Low energy costs attract residents who can then support culturally and socially relevant businesses. These socially and culturally relevant businesses and scenes would then be attractive to universities and other institutions which benefit the government with research opportunities. This perpetuates the cycle of a sustainable neighborhood system.

A relevant example of this is Eco-Viikki in Finland. This neighborhood was designed with intentions to create a sustainable, healthy, and comfortable living environment with solutions that save energy and reduce the amount of waste generated (Sitra, 2011). Apartment buildings use solar energy technology—both thermal and photovoltaic to keep energy costs low. They house university buildings, experimental farming, research institutes, and businesses. Eco-Vikki is rapidly growing, has successfully addressed the five pillars of sustainability, and is a pioneer in the market of sustainable urban neighborhoods.

2.4.4 A Life Cycle Approach

The final sustainable principle is to ensure a *life cycle approach* to planning. Be it through legislation or physical components, the mark of good decision-making for a sustainable system is a project's ability to sustain itself through its entire life cycle. Instilling systems that can remain flexible throughout their years and adapt to the changing needs of users is key (Fig. 2.12). It also must be considered that components within a project might have different life spans. A creditable sustainable project will ensure all components are seen through *cradle to cradle*. With this in mind, the life cycle approach also promotes as lengthy a life span as possible. Facebook marketplace as an example is an online marketplace that allows people to easily resell or give away used or unwanted items. Users have the option to tailor their searches geographically, categorically, and by price to find exactly what they are searching for. The ease, accessibility, and profitability of this platform and the many others like it incentivize users to recycle giving household items longer life spans.

2.5 Ecological and Carbon Footprint

The information age we face has allowed for the quantification of data as an application of technology. With sustainability being such a broad, all-encompassing term, there are numerous indices available to measure the sustainability of specific individuals, lifestyles, nations, companies, and more. The concepts of ecological and carbon footprints both take advantage of available data to establish the impact of anthropogenic activity on natural resources and ecosystems.

2.5.1 Ecological Footprint

The concept of the *ecological footprint* attempts to calculate the amount of land required to produce goods and assimilate waste generated by an individual, population, or product (World Wildlife Foundation, n.d.Global Footprint Network, n.d.-a). It accounts for built-up land, carbon emission levels, cropland, fishing grounds, forest land, and grazing land and is measured in global hectares (Gha) (Sarkodie, 2021). More specifically, ecological footprint per person is calculated by dividing a nation's entire ecological footprint by its population. In order to meet the needs of

Item	Contribution to Flexibility
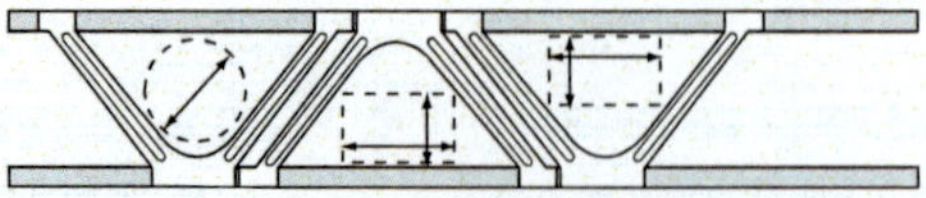 Space Joist	• Allows wide spans and as a result eliminate the need for additional support walls • The open web allows for easy 'fishing' of utility conduits
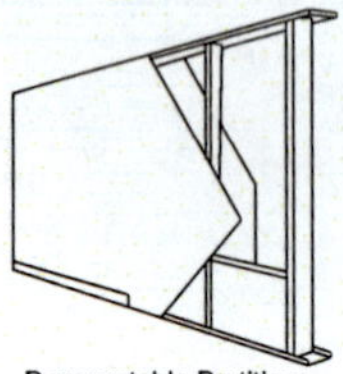 Demountable Partitions	• Allow easy placement of partitions anywhere on a floor and their dismantle and reuse
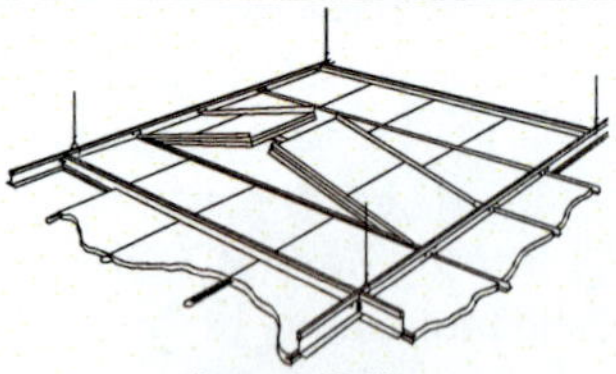 Suspended Ceiling	• Allow quick covering of exposed joists and easy access to condiuts for their future alteration
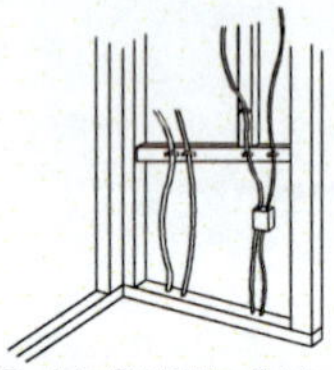 Flexibile Cold Water Tubing	• Allow easier instalation and replacement of plumbing fixtures
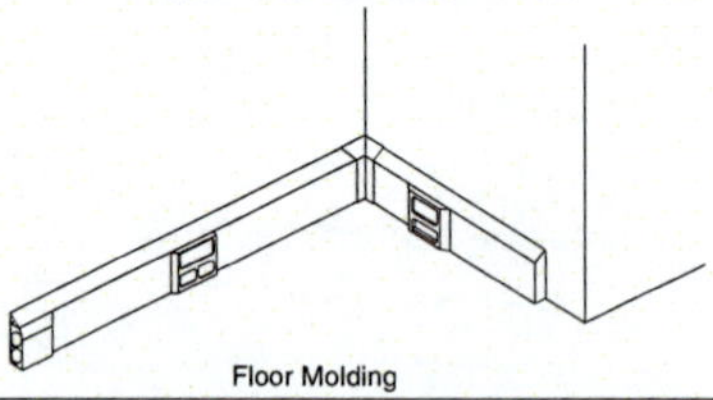 Floor Molding	• Allow flexiblity in yhr installation of recepticals and avoids the need to pass wires through the wall

Fig. 2.12 Using specialized products can help occupants maintain flexibility throughout their residency and adapt their dwellings to their changing needs

the global population, the world's ecological footprint must be equal to the Earth's biocapacity—its ability to produce needed resources and absorb waste—which is 1.7 Gha/person. In 2017, however, the global average ecological footprint was 2.8 Gha/person, with some nations such as Qatar having averages of up to 14.3 Gha/person (Global Footprint Network, n.d.-b). As evident, the ecological footprint of the Earth is presently functioning at a deficit, a path that cannot be sustained by present resources.

In order to reduce the global ecological footprint, action must be taken at both individual and institutional levels in some of the many ways that have been and will be discussed in this book. Reducing food waste, protecting natural ecosystems, maximizing energy efficiency, and supporting policies that enforce these initiatives are all important. Urbanization can either increase or decrease ecological footprints of citizens based on a variety of factors. Because of this, as trends of urbanization increase, integrating sustainable practices into the lives of city dwellers through technology and other means will be vital to reach equilibrium between ecological footprint and biocapacity.

2.5.2 Carbon Footprint and Offsets

Another index of sustainability commonly used is the *carbon footprint*. It is the amount of carbon released into the atmosphere (measured in tons) by an industry, activity, or vehicle. It is technically included within the measurement of an ecological footprint; however, because it is rapidly growing, makes up 60 percent of the global ecological footprint, and acts as a GHG contributing to climate change and its accompaniments, carbon has an index of its own (Global Footprint Network, n.d.-a). As mentioned earlier in this chapter, anthropogenically produced carbon in the atmosphere is largely a result of the burning of fossil fuels, such as coal, and those used for vehicle fuel and can be reduced by adopting clean and active transportation methods and clean energy sources, among other strategies. Similar to ecological footprint, the emissions of carbon are disproportionately produced in mass by only a few countries. China, the United States, and India contribute 49.4 percent of carbon into the atmosphere yet make up only 41.9 percent of the world's population (Global Footprint Network, n.d.-b).

Carbon emissions and urbanization have a similar relationship to that of the ecological footprint and urbanization. However, it is worth noting that study results vary widely based on criteria included. On the one hand, urbanization has historically required land and natural resources to be paved over for city dwellers. This eliminates many natural carbon sinks which are environments that absorb carbon dioxide from the atmosphere such as grasslands, agricultural land, peat bogs, and rain forests (Konyn, 2021). On the other hand, research such as that performed by Mirzabaev et al. (2020) in Austria shows that of rural, semi-urban, and urban dwellers, those living in urban environments have the lowest carbon footprint. This is owing to smaller housing; high density levels and availability of public transport however largely apply to high-income countries. Urbanization in low- and middle-income countries often results in higher carbon emissions per capita (Mirzabaev et al., 2020). In the coming years, 95 percent of urban expansion will take place in the developing world, often associated with low- and middle-income nations (UN, n.d.-a). This shows that while urban environments have the potential to lower carbon footprint, they presently do not in most of the world, hence the need for clean technology and other carbon-reducing technologies.

Carbon offsets are highly advertised and now used as a standard approach when a carbon-emitting activity is unavoidable under present circumstances. An individual or company purchases a carbon credit which in turn pledges another initiative to either lower emissions or sequester an equivalent amount of carbon so that the initial company can emit *net zero*. This approach is commonly used when people must travel via plane to offset their emissions but can also be used by governments and corporations. The IPCC has most recently announced that carbon emissions need to be cut by 45 percent from 2010 levels by 2030 and net-zero carbon emissions must be achieved by 2050 to mitigate climate change (IPCC, 2018). Carbon offsets, while they act as a present-day patchwork solution, are not contributing to the achievement of carbon reduction goals and allow for fossil fuel-dependent corporations to circumvent the adoption of clean technology.

2.6 Tomorrow's Sustainable World

As outlined above, be it philosophical or technical, the knowledge of how to achieve sustainable development and lifestyles is readily available. The pillars of sustainable development, the principles of sustainable systems, and their digital accompaniments are tools that can be used to achieve sustainability goals such as those outlined by the UN. What lies between today and a sustainable future is the implementation and actions of these holistic approaches. It needs to be however told that the most contemporary impending issues society faces today are intensified due to the unequal disparities of wealth both between and within nations. While technology holds potential to help solve many of these problems, lifestyle and culture changes must readily be made alongside technological development. Ahmed et al. (2020) highlights the importance of human capital based on education on influencing an individual's environmental awareness and pro-environment behavior such as using renewable energy, recycling, and acknowledgement of climate change and its consequences. Therefore, education is one of the most important tools in fighting climate change and implementing sustainable practices. Sustainable and education technologies are at the ready and can be used to expedite achieving set goals.

Questions for a Follow-Up Discussion

1. In which domains urgent measures need to be taken to lessen the climate emergency the world is facing?
2. Which of the social challenges the world is facing (e.g., climate change, aging) can be helped by using digital technologies?
3. How can digital technologies assist in food production and distribution?
4. What is the common definition of sustainability? Do you think it needs to be amended to meet contemporary social challenges?
5. Draw a link between the key pillars of sustainability and digital technologies.

References

Ahmed, Z., Ali, S., & Wasif, Z. M. (2020). Linking urbanization, human capital and the ecological footprint in G7 countries: An empirical analysis. *Sustainable Cities and Society, 55*, 102064. https://doi.org/10.1016/j.scs.2020.102064. Accessed April 10, 2022.

Alalouch, C., Bouzguenda, I., & Fava, N. (2019). Toward smart sustainable cities: A review on the role of digital citizen participation could play in advancing social sustainability. *Sustainable Cities and Society, 50*. https://doi.org/10.1016/j.scs.2019.101627. Accessed April 10, 2022.

Arnett, H. (2017). The challenges in quantifying social sustainability at a neighborhood level. *Cities & Health, 1*(2), 139–140. https://doi.org/10.1080/23748834.2018.1460104. Accessed April 10, 2022.

Bajya, P., Choudhary, M., Kaswan, S., Kaswan, V., & Kumar, P. (2019). *Green production strategies*. Encyclopedia of Food Security and Sustainability. Elsevier. https://doi.org/10.1016/B978-0-08-100596-5.22292-0. Accessed April 10, 2022.

Campbell, A., & Tindall, A. (2010). "Water security: National and global issues". *US Geological Survey*. https://pubs.usgs.gov/fs/2010/3106/pdf/FS10-3106.pdf. Accessed 10 April 2022.

Canadian Commission for UNESCO. (2020). "Connecting Culture and Sustainability. https://en.ccunesco.ca/blog/2020/1/connecting-culture-and-sustainability. Accessed April 10, 2022

Caradonna, L. (2018). "Part III historiography of sustainability". *Routledge International Handbooks*. Ch in Routledge Handbook of the History of Sustainability. https://books.google.ca/books?hl=en&lr=&id=CoI6DwAAQBAJ&oi=fnd&pg=PT14&dq=history+of+sustainability&ots=1LG7U9JxcL&sig=XyrBQclQT8gOES2-SEjGv8w3tus#v=onepage&q=history%20of%20sustainability&f=false. Accessed April 10, 2022.

Contento, A. (2018). "Water scarcity: The Most understated global security risk". *Harvard Law School National Security Journal*. https://harvardnsj.org/2018/05/water-scarcity-the-most-understated-global-security-risk/#_edn11. Accessed April 10, 2022.

FAO. (2021). "The State of Food Security and Nutrition in the World 2021". *Food and Agriculture Organization of the United Nations.*. https://www.fao.org/state-of-food-security-nutrition. Accessed April 10, 2022.

Friedman, A. (2007). *Sustainable residential developments: Design principles for green neighbourhoods*. McGraw-Hill.

Global Footprint Network. (n.d.-a). "Climate change". https://www.footprintnetwork.org/our-work/climate-change/. Accessed April 10, 2022.

Global Footprint Network. (n.d.-b). "Data". https://data.footprintnetwork.org/#/. Accessed April 10, 2022.

Global Water Security. (n.d.). "Water security solutions centre". https://globalwatersecurity.org/. Accessed April 10, 2022.

IPCC. (2018). "Summary for policy makers". *Special report: Global warming of 1.5°C*. Chap 00. https://www.ipcc.ch/sr15/chapter/spm/. Accessed April 10, 2022.

IPCC. (n.d.). "About the IPCC". https://www.ipcc.ch/about/. Accessed April 11, 2022.

Konyn, C. (2021). "What are carbon sinks?". Earth.Org. https://earth.org/carbon-sinks/. Accessed April 10, 2022.

Merriam-Webster. (n.d.). "Economy". *Merriam-Webster.com Dictionary*. https://www.merriam-webster.com/dictionary/economy. Accessed April 10, 2022.

Mirzabaev, A., Muñoz, P., & Zwick, S. (2020). The impact of urbanization on Austria's carbon footprint. *Journal of Cleaner Production, 263*, 121326. https://doi.org/10.1016/j.jclepro.2020.121326. Accessed April 10, 2022.

Pavid, K. (n.d.). "What is the Anthropocene and why does it matter?". Natural History Museum. https://www.nhm.ac.uk/discover/what-is-the-anthropocene.html. Accessed April 10, 2022.

Pfeffer, J. (2010). *Building sustainable organizations: The human factor*. Stanford University Graduate School of Business. https://papers.ssrn.com/sol3/papers.cfm?abstract_id=1545977. Accessed April 10, 2022

Sarkodie, S. A. (2021). Environmental performance, biocapacity, carbon & ecological footprint of nations: Drivers, trends and mitigation options. *Science of The Total Environment., 751*, 141912. https://doi.org/10.1016/j.scitotenv.2020.141912. Accessed April 10, 2022.

Schraad-Tischler, D., & Seelkopf, L. (2015). "Concept and mMethodology – Sustainable gGovernance iIndicators 2015". Sustainable Governance Indicators. https://www.sgi-network.org/docs/2020/basics/SGI_Concept.pdf. Accessed April 11, 2022.

Second Harvest. (n.d.). Second Harvest food rescue.. https://secondharvest.ca/foodrescue/app/terms-of-use/. Accessed April 10, 2022.

SGI. (n.d.). "About the SGI". *Sustainable Governance Indicators.*. https://www.sgi-network.org/2020/About. Accessed April 10, 2022.

Sitra. (2011). "World-class sustainable solutions from FinlandFinland". Finnish Innovation Fund Sitra, VTT Technical Research Centre of Finland, Tekes. https://www.sitra.fi/en/publications/world-class-sustainable-solutions-finland/. Accessed April 10, 2022.

Sitra. (2020). "Connecting food businesses and charities to capture surplus food". *The Finnish Innovation Fund Sitra*. https://www.sitra.fi/en/cases/connecting-food-businesses-and-charities-to-capture-surplus-food/. Accessed April 10, 2022.

Stobierskie, T. (2021). "4 impactful sustainable business practices that can makes a difference". *Harvard Business School Online.*. https://online.hbs.edu/blog/post/sustainable-business-practices. Accessed April 10, 2022.

UN. (1972). "Report of the united nations conference on the human environment". United Nations.. https://www.un.org/ga/search/view_doc.asp?symbol=A/CONF.48/14/REV.1. Accessed April 10, 2022.

UN. (n.d.-a). "11: Sustainable cities and communities". https://www.un.org/sustainabledevelopment/cities/. Accessed April 10, 2022.

UN. (n.d.-b). "Sustainability". https://www.un.org/en/academic-impact/sustainability#:~:text=In%201987%2C%20the%20United%20Nations,development%20needs%2C%20but%20with%20the. Accessed April 10, 2022.

UN. (n.d.-c). "The 17 goals". *United Nations Department of Economic and Social Affairs - Sustainable Development.* https://unric.org/en/united-nations-sustainable-development-goals/#:~:text=The%20SDGs%20are%20integrated%E2%80%94that,for%20those%20furthest%20behind%20first. Accessed April 10, 2022.

UN. (n.d.-d). "What is climate change?". https://www.un.org/en/climatechange/what-is-climate-change. Accessed April 10, 2022.

University of Gävle. (2018). "Economic sustainability". https://www.hig.se/Ext/En/University-of-Gavle/About-the-University/Environmental-Work/What-is-sustainable-development-at-HiG/Economic-sustainability.html. Accessed April 10, 2022.

University of Toronto. (2017). *Resilience planning guide*. John H. Daniels Faculty of Architecture. Landscape and Design.

Water Science School. (2019). "How Much Water is There on Earth?". *U.S. Geological Survey.* https://www.usgs.gov/special-topics/water-science-school/science/how-much-water-there-earth#:~:text=a%20watery%20place.-,But%20just%20how%20much%20water%20exists%20on%2C%20in%2C%20and%20above,percent%20of%20all%20Earth's%20water. Accessed April 10, 2022.

Wheeler, S. (2004). "Planning for sustainability: Creating livable equitable and ecological communities". https://books.google.ca/books/about/Planning_for_Sustainability.html?id=rtghgHKV6pwC&source=kp_book_description&redir_esc=y. Accessed April 10, 2022.

WHO. (n.d.). "Air pollution". *World Health Organization*. https://www.who.int/health-topics/air-pollution#tab=tab_2. Accessed April 10, 2022.

Woon, J. (2018). "Technology and governance in Singapore's smart nation initiative". *ASH Center for democratic governance and innovation.* Harvard Kennedy School.. https://ash.harvard.edu/publications/technology-and-governance-singapore%E2%80%99s-smart-nation-initiative. Accessed April 10 2022.

World Wildlife Foundation. (n.d.). "Ecological Footprint". https://wwf.panda.org/discover/knowledge_hub/teacher_resources/webfieldtrips/ecological_balance/eco_footprint/. Accessed April 10, 2022.

Chapter 3
Efficient, Safe, and Comfortable Mobility

The merger of digital infrastructure and urban mobility networks has the potential to redefine interurban transportation. When traffic sensors, streetlights, and bus lines all connect via digital signals and the Internet of things (IoT), a smart transportation network that maximizes efficiency emerges. Concurrently, a rise in rideshare apps like Uber and Lyft has changed the way people navigate cities by reducing the need to own a car. By creating more efficient public transit networks through an array of interconnected sensors and promoting ridesharing or car-sharing forms of transit, smart cities can reduce the presence of vehicles in cities which would in turn reduce carbon emissions and provide vast amounts of reclaimable space for urban improvement. This chapter introduces aspects related to the integration of mobility and digital means. It investigates the relations between digital means and mobility. Shared means of transportation and active mobility will be among the subjects explored.

3.1 Setting the Stage

Poor or inefficient mobility is a multifaceted issue that can be observed in many cities around the world. Car-centric city planning and design have led to significant negative consequences that affect the environment and accelerate climate change, as well as an individual's ability to navigate urban spaces in a peaceful, efficient, and productive manner. Air and noise pollution caused by excessive vehicle use can have severe health consequences for people living in cities and affect their ability to function normally (Wagner, 2018; World Health Organization: Regional Office for Europe, 2011). A study by Amin et al. (2017) in Montreal illustrates how the expansion of road networks contributes to poor air quality, while other studies elaborate on the connection between noisy streets and a variety of health issues. The effects of congestion and traffic on productivity are also mentioned, taking into account the

A. Friedman, *The Sustainable Digital City*,
https://doi.org/10.1007/978-3-031-25488-8_3

estimated total economic loss in the United States as a result of extended car commute times (Amin et al., 2017). By looking at the effects of these issues, cities can re-evaluate and reorganize transportation infrastructures to respond more effectively to present and future environmental impacts at local and global scales and improve the well-being of urban dwellers. Some of the negative consequences of poor transit planning will be outlined below.

3.1.1 Environment and Climate Change

Tackling climate change is a challenge that will affect people for generations to come. As per the report by the U.S. Department of Transportation (DOT) (2010) Center for Climate Change and Environmental Forecasting, greenhouse gas (GHG) emissions must be reduced significantly to curb the rise of global temperatures and limit warming between 2.0 °C and 2.4 °C (3.6 °F to 4.3 °F). Comparatively, the Paris Agreement signed by 191 countries as of July 2021 outlines that warming should be reduced to a minimum of below 2 °C (3.6 °F), preferably closer to 1.5 °C (2.7 °F) (United Nations, n.d.). More recent studies, however, strongly suggest that this is not enough to avoid ecological disaster. According to the Intergovernmental Panel on Climate Change (IPCC) (2021), reducing warming to1.5 °C (2.7 °F), despite not completely eradicating the likelihood of drastic environmental damage and severe weather events, allows enough room for human adaption and possible ecological regeneration. In recent years, countries around the world are recording their hottest days and months on record and experiencing severe environmental disasters at an abnormally frequent rate. July 2021 was officially declared the hottest month globally on record by the National Oceanic and Atmospheric Administration (NOAA) National Centers for Environmental Information (2021). The increasingly rapid production of GHG emissions is contributing to the rise of global temperatures, with a considerable portion being produced by the transportation sector as illustrated in Fig. 3.1. It is reported that transportation accounts for about 29 percent of total emissions produced in the United States and 5 percent globally (DOT 2010). Furthermore, over 90 percent of fuel used to run cars is petroleum based, which includes diesel and gasoline (US Environmental Protection Agency, 2019).

3.1.2 Pollution Caused by Road Traffic

An overdependence on fossil fuel-powered cars paired with further development of road infrastructures and inefficient management of traffic in cities results in elevated levels of air and noise pollution which can have variable health consequences for city dwellers. The most common pollutants released by cars are nitrogen oxides in addition to particulate matter, carbon monoxide, volatile organic compounds, and

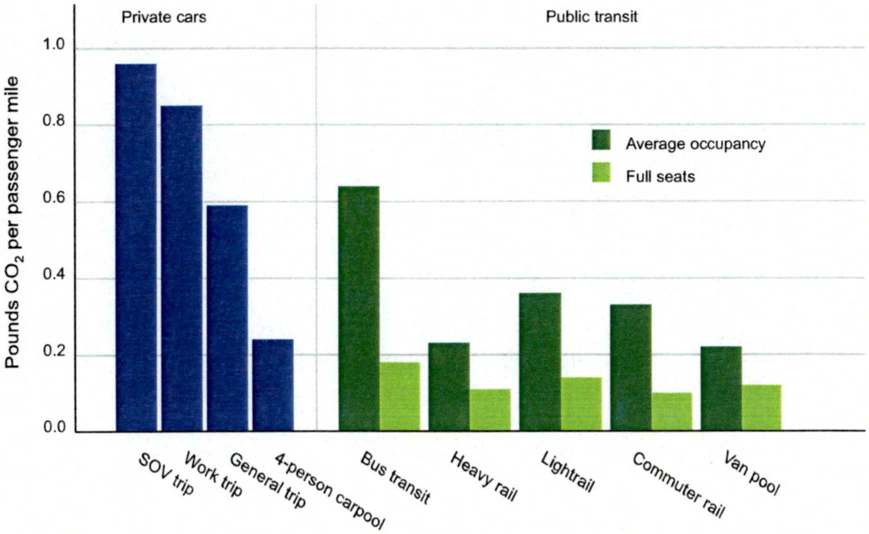

Fig. 3.1 Tailpipe emissions of common transport modes per passenger mile

sulfur dioxide (Amin et al., 2017). These compounds and particles, when inhaled over an extended period, can have serious health consequences. In 2014, the UN estimated that over seven million people globally died prematurely due to poor air quality (Bouton et al., 2015).

It is equally important to note the relationship between road network developments, capacity, and use. Social scientists have observed that when new roads are built, traffic gradually increases due to a phenomenon known as induced demand (Mann, 2014). This means that urban projects seeking to expand road networks and car capacity will very likely influence and even prompt people to drive over other methods of transportation creating environments that are increasingly unhealthy and polluted. This can be seen in the expectant increase in car sales from 70 million in 2010 to an expected 125 million by 2025 (Bouton et al., 2015). In the study that was noted above in Montreal, Canada, Amin et al. (2017) found that following the construction of Highway 25, which was completed in 2007 and opened in 2011, the levels of nitrogen oxide were increased in areas closest to the new highway as a result of increased traffic. The opposite happened during the COVID-19 lockdowns when people did not travel leading to sharp decrease in emissions. Finding ways to reduce car use and promote other forms of transit while maintaining people's ability to move around efficiently using digital means should be a priority for cities looking to improve their air quality and preemptively strengthen the health of their residents.

Aside from air pollution, cars also create environments that are disruptive to the senses. Noise pollution is an often overlooked form of pollution in cities that, overtime, can have a considerable impact on the physical and mental health of urban dwellers, adversely affecting their ability to function. Traffic is one of the primary sources of elevated noise levels in the city. Generally, urban dwellers will hear sounds periodically at a collective level of 85 decibels and about 60 decibels sustained and extended throughout the day (Wagner, 2018). Although uncertain, some studies have related these loud environments to cardiovascular diseases and other issues such as an increase in blood pressure and heart rate (Wagner, 2018; World Health Organization: Regional Office for Europe, 2011). Children are particularly affected by constant loud noises. Multiple studies have shown that chronic noise exposure can affect children's memory, ability to focus and retain information, reading, language processing, and overall development (World Health Organization: Regional Office for Europe, 2011). Adults at work may also have trouble focusing on a given task or working efficiently due to elevated noise levels and in some cases develop tinnitus, which only exacerbates the effects mentioned prior (World Health Organization: Regional Office for Europe, 2011).

Elevated urban noise levels caused by road traffic are also responsible for sleep disturbance and/or deprivation and noise annoyance. As outlined in the WHO Regional Office for Europe (2011) report, the body will respond to noise while sleeping, part of naturally occurring and vital sleep responses. Consequently, when exposed to chronic noise, these reactions are triggered too often and can cause discomfort among other health issues. An individual experiencing chronic noise above 40 dB at night will begin to feel adverse effects due to a lack of a restful sleep. These effects such as an inability to perform tasks in the daytime, decrease in creativity and logical decision-making, and general fatigue are far worsened when nighttime noise rises above 55 dB (World Health Organization: Regional Office for Europe, 2011).

3.1.3 Efficiency of Traffic and Productivity

Congestion occurs when roads become saturated, that is, when the number of cars on the road exceeds its capacity (Harriet et al., 2013). Trends of increasing vehicle sales are still being observed in the United States. In fact, the number of vehicle miles traveled (VMT) by light-duty motor vehicles including passenger cars and light-duty trucks increased by 48 percent from 1990 to 2019, due to confluence of factors including population and economic growth, urban sprawl, and periods of low fuel prices (EPA 2019). Figure 3.2 illustrates the relations between residential density and VMT. This sustained increase in vehicles' use impacts not only the time drivers spend on the road due to road saturation, delaying their trips by a considerable amount, but also the distribution of goods and services that business and their customers rely on (Harriet et al., 2013). With an estimated five billion people expected to be living in urban areas by 2030 and, as mentioned previously, a steady increase in the sale and use of vehicles, it will be impossible to support driving as the primary form of urban mobility within the present transportation frameworks (UN, n.d.; Bouton et al., 2015). Transportation infrastructures and networks in many cities around the world are or will soon be incapable of managing the number of drivers on the road, resulting in seemingly inescapable congestion throughout the

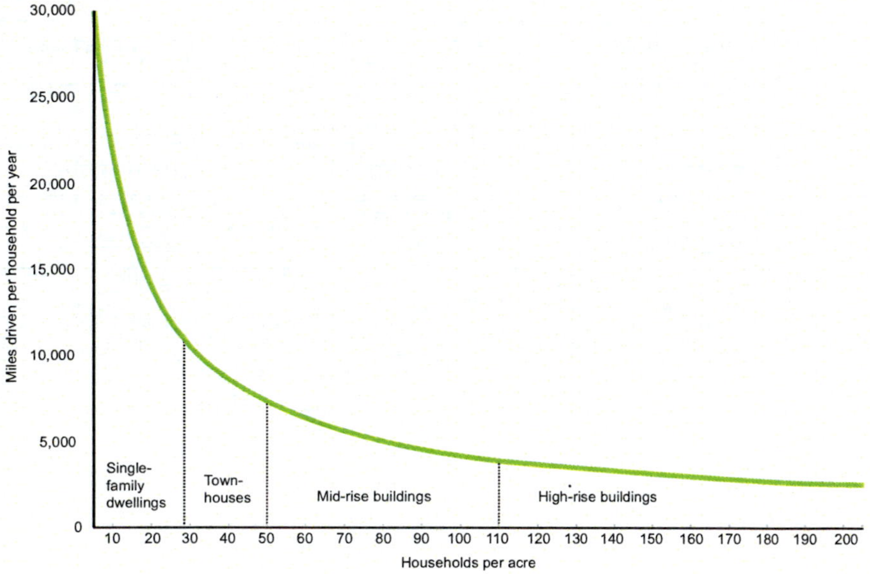

Fig. 3.2 Residents living in the US communities with higher densities tend to drive three times less than those who live in neighborhoods with single-family homes

day which will ultimately result in a loss of productivity and economic opportunity. A study conducted by transport data company INRIX across multiple global cities in the United States found that traffic congestion resulted in an average loss of hundreds of hours every year per driver, reaching as high as 164 h in Boston (Fleming, 2019). The data company also estimated that $87 billion USD is lost per year due to a loss of productivity associated with congestion (Fleming, 2019). Shifting focus away from car-centric transportation networks could prove economically beneficial to cities. Now that there is a general understanding of the unsustainability of present transportation habits, the following sections will present an array of options aside from the privately owned vehicle, all of which are supported by the digital age.

3.2 A New Focus on Public Transportation and Ridesharing

Developments over the late twentieth and early twenty-first centuries have continued to favor private vehicles as the primary and most dependable method of transportation. It is becoming increasingly clear, however, that cities should shift more of their attention away from private car ownership toward other more efficient modes of mobility, such as public transit, cycling, and walking; these methods emit low to no carbon emissions. Public transit can be improved and made increasingly desirable to commuters by using GPS technologies, intelligent transportation system (ITS), and real-time data as done in cities like Montreal and Copenhagen. Furthermore, the recent rise of ridesharing and car-sharing programs that provide more flexibility to urban transportation is another area worth discussing in terms of easing road congestion. Bike-sharing is also shown to be a viable solution to problems of mobility in many cities across North America and Europe, while car-sharing can more effectively utilize the available seats in every car (Figs. 3.3 and 3.4). Consequently, the number of cars on the road can be lowered as can the intensity of congestion and levels of GHG emissions. Lastly, in this section, transport network companies (TNCs) like Uber and Lyft are considered, and their impacts on road congestion are discussed, presenting a gray area in ridesharing that may not always prove beneficial in decreasing car usage in cities overall.

3.2.1 *Improving Public Transit Using Technologies*

For years, cities have been publishing daily bus schedules for riders to best plan their routes. While these schedules aim to be as accurate as possible, they are often incapable of accounting for unexpected travel delays because of construction, detours, accidents, and harsh weather especially in the winter months. The disconnect between the planned bus schedule and the real-time bus schedule can create problems for people who rely on public transport to get around and results in a lack of trust in said services. It also makes buses slower and less reliable than privately owned vehicles, discouraging those attempting to make the switch from private to

Fig. 3.3 Car-sharing can utilize the available seats in every car more effectively

Fig. 3.4 A station for car-sharing electric vehicles in Berlin

public transportation methods. With modern technology, however, many transit systems have employed GPS and IoT technologies—as introduced in Chap. 1—that track and communicate bus locations to provide riders with the most accurate and reliable bus location information. It can now be clear to users when a given bus, train, metro, or subway will arrive at or depart from their stop (Fig. 3.5). This helps to reduce travel time and increase convenience as riders know exactly when they will need to be at their stop in the case that there is a delay. The Société de Transport de Montréal (STM) in Montreal, for example, offers many ways for riders to find out when their bus will arrive and/or depart in real time using GPS and odometer data within 60 minutes of the original planned arrival and/or departure (Société de Transport de Montréal, n.d.). Riders can access updated schedule information from the STM website, for example. Similarly, the downloadable "Transit" app allows users to view updated schedules and track their bus's live locations (Fig. 3.6). Lastly, if a potential rider lacks access to the Internet, there are phone numbers posted at bus stations that riders can text which respond automatically with a real-time schedule according to specified stop and bus numbers.

In other cities, smart traffic systems are integrated with road infrastructure to improve the efficiency and desirability of cycling and public transit. These technologies are notably employed in Copenhagen, Denmark, where the city—already known for its high active mobility and public transit use—is making increased efforts to reduce bus travel times by 5–20 percent through the integration of intelligent traffic signals (ITS) alongside their goal of becoming carbon neutral by 2025 (Davies, 2016; Sustainia, 2018). The ITS system works by receiving real-time

Fig. 3.5 With modern technology, many transit systems have employed GPS and IoT technologies that track and communicate bus locations to provide riders with the most accurate and reliable information

information from buses on the road, like how many people are currently aboard, their position, and if the bus is experiencing any delays. The traffic lights are then able to respond accordingly, adjusting lights to favor the passage of buses. This means that a light could stay green for 8–30 seconds longer, allowing a bus to pass through a busy intersection at rush hour (Davies, 2016). It is also worth noting that the system is equally beneficial to cyclists as well, expected to lower said travel times by ten percent (Osborne Clarke, 2016), and cars by reducing overall congestion and improving traffic flow (Sustainia, 2018). Through their implementation, ITS and other similar technologies can promote public transit and active mobility in cities and reduce carbon emissions, noise levels, and the use of private vehicles. Prioritizing and improving public transit using these technologies is also beneficial to multiple groups of people, including low-income residents and those with impaired mobility, and has rippling social and economic effects across the urban fabric (Friedman, 2014). By providing accessible and timely public transit, people will be able to rely on the services to get them to where they need to go without wasting time.

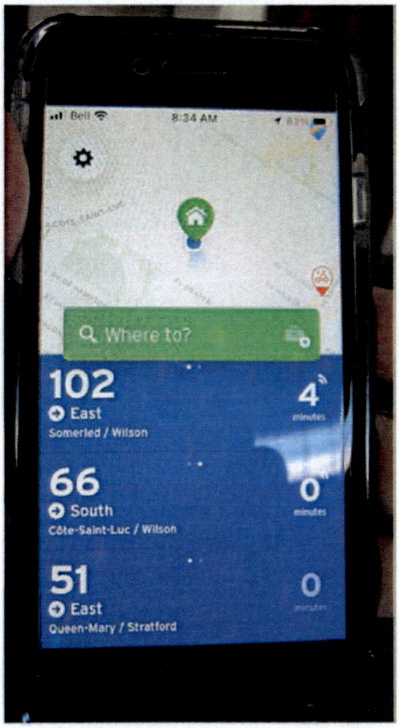

Fig. 3.6 Many public transit authorities introduced applications like the Montreal's Transit App which let users know in real time about bus or train time of arrival to a stop

3.2.2 Ridesharing and Car Ownership

The recent introduction of ridesharing programs in cities is also worth acknowledging and is a legitimate substitute for the privately owned car for urban dwellers. Both car- and bike-sharing programs are quite accessible, available at most if not all times of the day and night throughout the year via cellular applications. In particular, bike-sharing programs have seen an interesting trajectory in cities in North America in the last decade or so. Shared-use bike programs enable people to rent a bicycle at ports stationed around core parts of the city at any time of the day without having to make a prior reservation, allowing a great amount of flexibility to users (Shaheen et al., 2013). At the end of a trip, the bike is returned to the original port or another one near the user's destination ready to be used by another rider (Fig. 3.7). In a survey of bike-share riders in Washington DC, 70 percent indicated that they chose bike-sharing as their preferred mode of transport as it was the most convenient method of getting to where they needed to go in the city (Sobolevsky et al., 2018). Vancouver, Canada, for example, has a digitally enabled public bike-sharing system entitled Mobi (n.d.) that was launched in the summer of 2016. Users register by creating an account, selecting a plan, and entering payment information either online or through the Mobi app. With this account, users can use the app to see where bikes are available around the city and use a code provided by the app to

Fig. 3.7 Bike-share stations in cities around the world

unlock the bike and use as though it were their own. At the end of a Mobi bike trip, riders dock their bike at a designated Mobi station, and the duration of the trip is noted and charged to the account (Mobi, n.d.).

The active mode of transport provides many benefits to the city and riders alike, which include reducing vehicular congestion on the road while promoting a healthy and environmentally friendly lifestyle, and has become a popular choice for urban dwellers looking to improve their own health and/or combat climate change (Midgley, 2009). Studies on the topic of carbon emission reduction as a result of bike-share programs have been conducted across multiple systems in North America. For example, Sobolevsky et al. (2018) found that Denver's B-Cycle program has saved 466.7 tons of CO2 emission, with 43 percent of B-Cycle rides replacing car trips, resulting in a 15,868 gallon (60,067 liter) decrease in gasoline. They also found a two percent to three percent reduction in traffic attributed to the use of the Capital Bikeshare program in Washington DC within its first 3 years of operation (2018). Studies have also been conducted that focus on the Citi Bike system in New York City launched in May 2013. Citi Bike, which is the biggest rideshare system in the United States with over 600 ports and 10,000 bicycles, has managed to equally impact carbon emissions in the city with each station estimated to save 0.2 to 2 metric tons of carbon dioxide per year depending on their location (Sobolevsky et al., 2018). Bike-sharing can also be used as an accessible solution to complement and resolve some of the shortcomings of public transportation; nodes and stops that may otherwise be spread far apart can be brought closer by strategically placing bike-share racks and parking docks close to public transit (Friedman, 2021). In many Dutch, Danish, and German cities, bike-share docks are arranged in this fashion (Pucher & Buehler, 2008). This would be an optimal solution, encouraging both the use of bike-share programs and public transit and essentially eliminating the need for a car in people's daily commutes around the city. In fact, studies have shown that bike-sharing complements and even increases the use of public transit (American Public Transportation Association, 2016).

On another hand, there are digitally managed car-sharing programs as mentioned above. Similarly to bike-sharing, car-sharing is a flexible way for citizens to get around the city without necessarily requiring ownership of said means of transport. Car-sharing programs do not directly eliminate GHG emissions in the transport sector but instead work to reduce emissions by reducing the number of cars on the road. In fact, every single car shared can replace up to 15 individually owned cars (Canada 2021). By reducing car ownership, seeing that car-sharing is an affordable and convenient option for many people living in cities, other road infrastructures like parking on the street or in a lot can be vacated to other less carbon-intensive modes of transportation (US Department of Transportation Federal Highway Administration, 2016).

In Canada, a person will spend on average an hour a day in their car travelling to work, and a vast majority choose to drive despite living within a distance reachable by other modes of transportation (Friedman, 2014). If people are less inclined to purchase a car—or not see a car as a necessity in urban mobility—they are more likely to consider taking a bike or a bus to get around for short-distance trips (D'Orso et al., 2020). Ridesharing programs can influence people to avoid buying cars and contribute to a decrease in road traffic. Doubly, car-sharing more effectively utilizes individual vehicles and the space offered to them. In studies conducted by UN Habitat III, a conference held in 2016 on Housing and Sustainable and Sustainable Urban Development, they found that vehicles remained parked 95 percent of the time and had an average occupancy rate of less than two people per vehicle while driving. In the United States specifically, the occupancy rate fell to 1.5 people per car (Soares et al., 2018). This information suggests that most cars are only being used to serve a limited function, and by maximizing the use of their space, fewer cars will be on the road. One solution cities could consider to influence people to car-share is a reduction of autoroute and/or inner-city lanes in tandem with an integration of reserved lanes for those carpooling or taking other forms of car ridesharing.

There are many viable car-sharing options. Car rental agencies such as Enterprise or Avon, for example, allow users to rent cars from branches around the world. Otherwise, peer-to-peer car-sharing cuts out the middleman and still reduces the number of cars on the roads. Other car-sharing programs such as Communauto are set up more similarly to bike-sharing options (Communauto Montreal, n.d.). Communauto provides access to hundreds of cars throughout Canada and France that can be rented by the minute, hour, or day. Online or through the app, drivers register their license with the company and select an annual plan and insurance rates. Following registration, users can log on to the app and can peruse the city to see if there is a car available near them either in that instant or for reservation at a later date (n.d.). Communauto uses IoT technology to reserve cars and allows the users' phones to lock or unlock the car at designated times at the beginning and end of trips. It also uses GPS technology to locate cars throughout cities. Programs such as these make car-sharing highly convenient and cheaper than privately owned vehicles. Sometimes, these rides can even be cheaper than public transit.

TNCs like Uber and Lyft have also capitalized on changing transportation habits and contributed to the growing number of rideshare programs in urban areas. It should be noted, however, that ride-hailing, which is one form of ridesharing, is not always effective in reducing congestion and may, in some cases, make it worse. Unlike traditional cabs, ridesharing programs, such as Uber, rely on an app to connect drivers to customers. Riders pre-enter payment information into the app and can order a car at a moment's notice by entering both pickup and drop-off locations into the app. A notable quirk of these programs is that both riders and drivers are connected to "rateable" profiles within the app. This holds both drivers and riders accountable for their behavior as a negative review can result in less drivers willing to pick up a poorly rated customer or less riders willing to accept a poorly rated driver. Uber also provides security features afforded by IoT and GPS technologies, such as the ability to "share my ride" with a friend or family member to confirm with a third party that the trip is completed safely. TNCs as they function today, however, tend to pull people away from public transport, walking, and cycling due to their accessibility and convenience that rival the other modes of mobility mentioned. Researchers studying the impacts of TNCs on congestion and public transit ridership found that congestion increased as TNC trips accounted for considerable amounts of miles traveled within cities, while public transit ridership decreased by about 8.9 percent (Diao et al., 2021). The findings illustrate that car-sharing programs may not always be a primary solution to congestion and elevated levels of air and noise pollution in urban areas.

3.3 Active Mobility

Cycling and walking are equally important methods of transportation to consider, especially in cities that want to reduce the overall use of cars and the resultant GHG emissions. A reallocation of space reserved from cars to cyclists and pedestrians is necessary for these modifications to occur. Many changes can be made that further distinguish pedestrian paths and bicycle lanes from vehicular traffic and improve the overall comfort and well-being of those choosing to walk or bike. Any changes made, however, should also keep vulnerable populations in mind to ensure that they also have access to active modes of mobility and the attached social, economic, and health benefits (Fig. 3.8). To make active transport safe and convenient, improvements can be made that better the connectivity and integration of sidewalks, bike lanes, and public transit nodes. This way commuters can better transition between active modes of mobility and public transit so that switching the two throughout daily commutes is convenient. This section will show how supporting active mobility through digital and other means can gradually reduce car dependency in cities.

3.3.1 *Improving the Safety and Comfort of Pedestrians and Cyclists*

It is necessary that the safety of all road users which includes pedestrians, cyclists, and drivers is considered when planning and/or adjusting road and other transit systems to favor active mobility. Cores of cities often lack suitable infrastructure and facilities for non-vehicular travel, especially bikes, which creates dangerous

Fig. 3.8 Singapore's public transit system was designed to keep vulnerable populations in mind to ensure that they also have access to active modes of transportation and their associated social, economic, and health benefits

environments for pedestrians and cyclists alike who are either forced to share the road with cars or are barred from them entirely. When city planners focus less on vehicles as the primary form of transportation and instead look to promote active mobility, urban areas can also become more enjoyable and safer overall for residents. However, to do so, systems and policies must be put in place and adopted to ensure favorable results. As discussed previously, expanding road infrastructures suited primarily or exclusively for vehicles only increases the number of cars on the road and does little to promote other forms of transportation. Rather, if we are looking to lower the number of cars, the space allocated to them needs to be reduced and given to safe urban design that supports pedestrians and cyclists (Fig. 3.9). This might entail the widening of sidewalks for pedestrians and integration of divided bike lanes, for example (Friedman, 2021). A reallocation of road space to cyclists specifically would also help to promote cycling as riders would feel a greater sense of security on their travels. Bike lanes should also include physical barriers that separate them from car traffic on main roads to further increase cyclists' safety (Friedman, 2021). Furthermore, the lane should be at least 1.5 meters (4.9 feet)

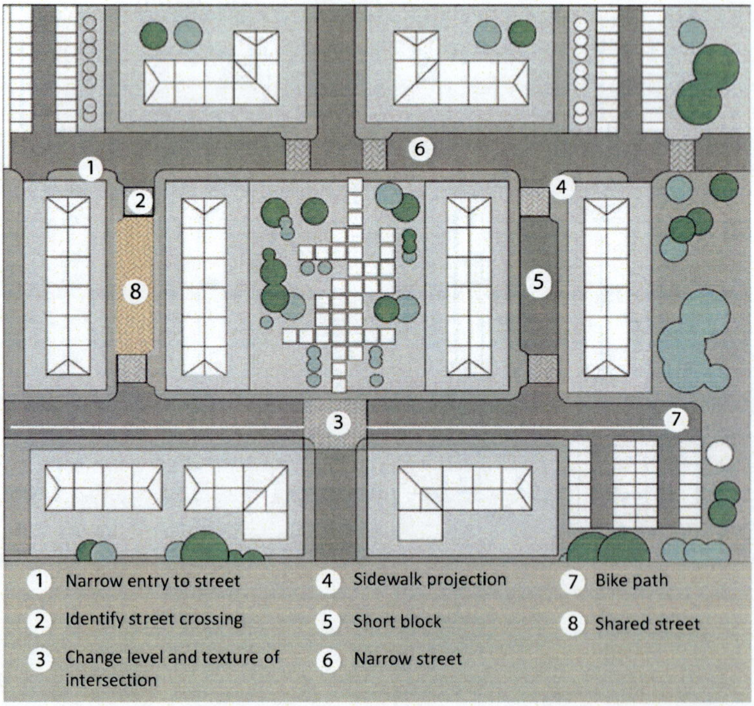

Fig. 3.9 The design of safer streets for pedestrians and cyclists includes traffic calming measures like changing the level and altering the texture of intersections and creating sidewalk extensions to narrow entry to a street

wide and be marked with signs and paint that clearly distinguishes the cyclist lane from the car lane(s) (NACTO, 2014).

City planners have a variety of digital tools at their disposal to make the least number of changes to routes that result in the most amount of people connected to adequate transport routes. Data generated from smartphones, for example, can provide information on the most traveled routes in the city. Traffic data can highlight the most congested areas—target populations—and can also show where the most accidents occur, vehicle to pedestrian, vehicle to bike, and others. GIS technologies allow for all this information and more, such as existing infrastructure and its age, to be considered in a single model. More on data and urban planning in the digital age can be found in Chap. 9.

Cycling and walking should also coordinate well with vehicular traffic as to avoid accidents. This can be done by lowering speeds in certain zones to 15 km/h or 30 km/h (10 mph or 18 mph), respectively, prioritizing cyclists and pedestrians on the road, congestion charging, and more (Friedman, 2014). Inversely, driving can be made more undesirable in cities as an indirect way of promoting active modes of mobility. This might entail, as mentioned above, the reduction of speed limits, an increase in parking charges, and changes to municipal policies that discourage private motorized vehicles, among other solutions (Friedman, 2014). Recall from Chap. 1 the transponders installed in private vehicles in Singapore that use GPS and IoT technologies to charge drivers when they drive in congested areas and/or at peak traffic times. Furthermore, more priority to pedestrians can be given in intersections to improve their safety. Crosswalks, for instance, can be made safer by becoming "smart." The new smart "zebra crossing" crosswalks developed in Britain, for example, use acoustic and visual alerts with the help of sensors, cameras, and other IoT technology (Liceras, 2022). It is a system made up of smart tiles that light up differently based on traffic circumstances and pedestrian requirements. The smart zebra crossing uses these technologies to determine when all the vehicles around a crosswalk are stopped and then light up the tiles indicating to the pedestrian that it is safe to cross. They can also turn red when something unsafe or unexpected has entered the roadway—like a ball and a fallen biker (2022). Lighting is another factor to consider and can likewise be made "smart." Sidewalks and other paths should be well lit to increase visibility at night and give people a greater sense of security, while adequate shading should be offered during the day—especially important to protect vulnerable people from excessive heat on warmer days which will only become more common due to climate change. A new lighting system in Amsterdam allows city officials to remotely change streetlight intensity based on area needs (Bravo, 2019). First responders also have access to these controls to change lighting to better respond to emergencies. In some areas of the city such as Atlas Park, citizens can access streetlight intensity control through an app (2019). All these measures work well together to improve the quality of life, safety, and comfort of cities, reduce road congestion, and boost economic activities in addition to increasing social interactions among urban dwellers (Friedman, 2021).

3.3.2 *Integrating Active Mobility into Public Transit*

In a report by the Transport Policy Advisory Services on behalf of the Federal Ministry for Economic Cooperation and Development in Germany, *Cycling-Inclusive Policy Development: A Handbook* (2009), ways to improve and support cycling as a viable and preferred method of transport are outlined. One of the noted points is to provide connections from bike routes to main public transportation stops and/or hubs that more effectively integrate biking into people's daily commutes (Transport Policy Advisory Services, 2009). This is known as multi-model planning, which considers a walking, cycling, and public transit as an interconnected system and would increase the convenience and efficiency of both active modes of mobility and public transit (Litman, 2017) (Fig. 3.10). Inversely, bike lanes might be extended further so that there are less bus stops along a line, encouraging people to take their bike to the bus or train. An integration of bike lanes could help to reduce strain on public transport systems so that they can continue to work efficiently while improving cycling conditions—the goal is to create a system in which bikes and public transport work together to distribute people across the city as opposed to fighting for the same space (Transport Policy Advisory Services, 2009).

Another point mentioned is to facilitate transfers between cycling and embarking on public transport (Transport Policy Advisory Services, 2009). Cycling is one of the most efficient ways to travel short to medium distances within the city. Consequently, when paired effectively with different modes of public transport like buses, metros, and trains, it can become extremely desirable for urban dwellers. Placing bike racks near transit nodes and stops and including spaces on public transit for cyclists to carry their bikes along with them make it so that they can transition between these differing modes of transportation with ease (Fig. 3.11). This also helps to prevent bicycle theft—a deterrent for many potential bicycle users (Transport Policy Advisory Services, 2009). Furthermore, it has been noted that an increase in active forms of mobility in urban areas improves the health and quality of life of people who chose to walk or bike, in addition to reducing vehicular congestion and air pollution on the road (Friedman, 2021).

3.4 Transportation for Thought

Transportation plays an immense role in our everyday lives. The way cities have grown in recent decades means that private vehicle transport is the easiest, safest, and fastest way for many to get around. Today, these transportation habits are changing. City dwellers and planners can see the environmental, economic, and social benefits to public and active methods of transportation. Aided by technological developments, there is much infrastructure being implemented by cities, private companies, and peer-to-peer collaboration which further a sustainable transport

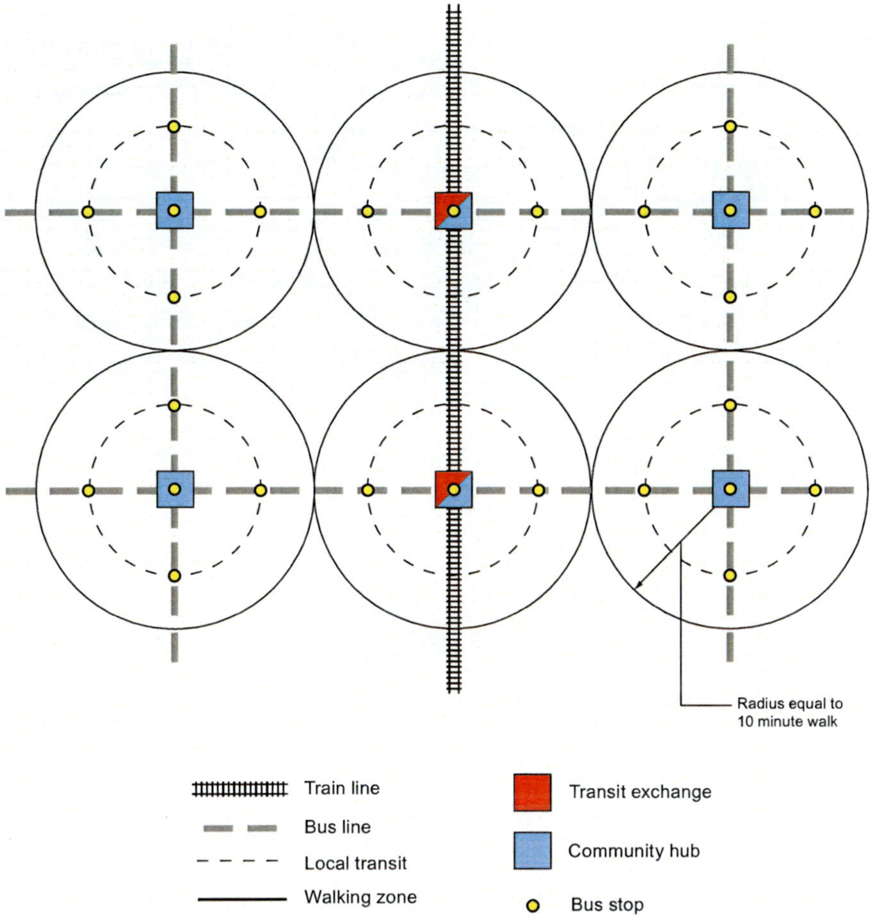

Fig. 3.10 Multi-modal planning considers walking, cycling, and public transit as an interconnected system that supports active mobility

agenda. As mentioned in this chapter, bike-share, car-share, and rideshare apps exist in many cities throughout the world. These sharing systems are cheap and accessible to city dwellers and tourists alike. They also contribute to a reduction in private vehicle traffic. Other technologies, such as smart lighting and crosswalks, and the collections and analysis of data work to keep active commuters as safe as possible. All these technologies support the shift away from private vehicle ownership. These changes mean fewer vehicles on the road, less congestion, and less noise pollution. They also contribute to fast, comfortable commutes, better air quality, and healthy, happy digital cities.

Fig. 3.11 Placing bike racks near transit nodes and stops and including spaces on public transit for cyclists to carry their bikes along with them helps transition between different modes of transportation with ease

Questions for a Follow-Up Discussion

1. What are the barriers associated with expanding use of public transit in major cities?
2. What is the potential that digital technologies hold to widen use of public transit in cities?
3. Is car-sharing the solution to traffic congestion in cities? How can it be made more efficient?
4. How can digital means make cycling and walking safer?
5. What future innovation do you foresee and suggest in the area of mobility in general and public transit in particular?

References

American Public Transportation Association. (2016). "Shared Mobility and the Transformation of Public Transit: Research Analysis." https://www.apta.com/wp-content/uploads/Resources/resources/reportsandpublications/Documents/APTA-Shared-Mobility.pdf. Accessed April 10, 2022.

Amin, S. R., Tamina, U., & Jimenez, L. A. (2017). Understanding air pollution from induced. Traffic during and after the Construction of a New Highway: Case Study of Highway 25 in Montreal.

Journal of Advanced Transportation, 2017, 1–14. https://doi.org/10.1155/2017/5161308. Accessed April 10, 2022.

Bouton, S., Knupfer, S. M., Mihov, I., & Swartz, S. (2015). "Urban mobility at a tipping point." *McKinsey Sustainability*. https://www.mckinsey.com/business-functions/sustainability/our-insights/urban-mobility-at-a-tipping-point. Accessed April 10, 2022.

Bravo, E. (2019). "Amsterdam launches a modular urban lighting system". *Tomorrow. City.* https://tomorrow.city/a/amsterdam-launches-a-modular-urban-lighting-system. Accessed April 26, 2022.

Commauto Montreal. (n.d.) "How it works". https://montreal.communauto.com/?lang=en&city=montreal. Accessed April 25, 2022.

D'Orso, G., Migliore, M., & Caminiti, D. (2020). "The environmental benefits of carsharing: The case study of Palermo." *Transportation Research Procedia, 48.* 2127–2139. https://doi.org/10.1016/j.trpro.2020.08.271. Accessed April 10, 2022.

Davies, A. (2016). "Copenhagen's new traffic lights recognize and favor cyclists." *Wired.* https://www.wired.com/2016/02/copenhagens-new-traffic-lights-recognize-and-favor-cyclists/. Accessed April 10, 2022.

Diao, M., Kong, H., & Zhao, J. (2021). Impacts of transportation network companies on urban mobility. *Nature Sustainability, 4*, 494–500. https://doi.org/10.1038/s41893-020-00678-z. Accessed April 10, 2022.

Fleming, S. (2019). "Traffic congestion cost the US economy nearly $87 billion in 2018." *World Economic Forum*. https://www.weforum.org/agenda/2019/03/traffic-congestion-cost-the-us-economy-nearly-87-billion-in-2018/. Accessed April 10, 2022.

Friedman, A. (2014). *Planning small and mid-sized towns; designing and retrofitting for sustainability*. Routledge.

Friedman, A. (2021). *Fundamentals of sustainable Urban Design*. Springer Publishing.

Harriet, T., Poku, K., & Emmanuel, A. K. (2013). An assessment of traffic congestion and its effect on Productivity in Urban Ghana. *International Journal of Business and Social Science, 4*(3), 225–234. https://ijbssnet.com/journals/Vol_4_No_3_March_2013/25.pdf. Accessed April 10, 2022

IPCC. (2021). "Special Report: Global Warming of 1.5 °C." https://www.ipcc.ch/sr15/. Accessed April 10, 2022.

Liceras, P. (2022). "A digital zebra crossing could be the future of our roads". *Tomorrow. City.* https://tomorrow.city/a/smart-zebra-crossing. Accessed April 26, 2022.

Litman, T. (2017). "Introduction to multi-modal transportation planning: Principles and practices." *Victoria Transport Policy Institute, Canada* https://www.vtpi.org/multimodal_planning.pdf. Accessed April 10 2022.

Mann, A. (2014) "What's up with that: Building bigger roads actually makes traffic. Worse." *Wired.* https://www.wired.com/2014/06/wuwt-traffic-induced-demand/. Accessed April 10, 2022.

Midgley, P. (2009). "The role of smart bike-sharing Systems in Urban Mobility." *Journeys, 2*(2). https://www.researchgate.net/publication/242235724_The_Role_of_Smart_Bike-sharing_Systems_in_Urban_Mobility. Accessed April 10, 2022.

Mobi. (n.d.). "How it works". https://www.mobibikes.ca/en/how-it-works. Accessed April 25, 2022.

National Association of City Transportation Officials (NACTO). (2014). "Conventional Bike. Lanes." https://nacto.org/publication/urban-bikeway-design-guide/bike-lanes/conventional-bike-lanes/#. Accessed April 2022.

National Centers for Environmental Information. (2021). "Global Climate Report – July 2021." https://www.ncdc.noaa.gov/sotc/global/202107. Accessed April 2022.

Osborne Clark. (2016). "The Danish capital city will spend 60 million kroner (£6.2m) to implement intelligent transport systems (ITS), including smart traffic lights that prioritise buses and bicycles." https://smartcities.osborneclarke.com/its/copenhagen-to-install-smart-traffic-lights/. Accessed April 2022.

Pucher, J., & Buehler, R. (2008). Making cycling irresistible: Lessons from The Netherlands, Denmark and Germany. *Transport Reviews, 28*(4), 495–528. https://doi.org/10.1080/01441640701806612. Accessed April 10, 2022.

Shaheen, S. A., Cohen, A., & Martin, E. (2013). Public Bikesharing in North America. *Transportation Research Record Journal of the Transportation Research Board., 2387*(1), 83–92. https://doi.org/10.3141/2387-10. Accessed April 10, 2022.

Soares, M. C. A., Marie de Salles Hue, N. P., Berssaneti, F. T., & Quintanilha, J. A. (2018). An Overview of Shared Mobility. *Sustainability., 10*(12). https://doi.org/10.3390/su10124342. Accessed April 10, 2022.

Sobolevsky, S., Levitskaya, E., Chan, E., Postel, M., & Kontokosta, C. (2018). *Impact of bike sharing in New York city*. A Report Submitted to Cornell University.

Société de Transport de Montréal. (n.d.). "iBUS – Real time." https://www.stm.info/en/about/major_projects/completed-major-projects/ibus-real-time. Accessed April 10, 2022.

Sustainia. (2018). "Copenhagen: Smart Traffic Signals Boost Cycling." *Global Opportunity Explorer*. https://goexplorer.org/copenhagen-smart-traffic-signals-boost-cycling/. Accessed April 10, 2022.

Transport Policy Advisory Services on behalf of the Federal Ministry for Economic Cooperation and Development. (2009). Cycling-Inclusive Policy Development: A Handbook. Utrecht. https://sutp.org/publications/cycling-inclusive-policy-development-a-handbook/. Accessed April 10, 2022.

U.S. Department of Transportation (DOT). (2010). Transportation's Role in Reducing U.S. Greenhouse Gas Emissions: Volume 1. Report to U.S. Congress. https://rosap.ntl.bts.gov/view/dot/17789. Accessed April 10, 2022.

U.S. Department of Transportation Federal Highway Administration. (2016). Shared Mobility: Current Practices and Guiding Principles. Washington, D.C. https://ops.fhwa.dot.gov/publications/fhwahop16022/index.htm. Accessed April 10, 2022.

UN. (n.d.). "11: Sustainable Cities and Communities". https://www.un.org/sustainabledevelopment/cities/. Accessed April 15, 2022.

United Nations. (n.d.). "The Paris Agreement." *UNFCCC Sites and platforms*. https://unfccc.int/process-and-meetings/the-paris-agreement/the-paris-agreement. Accessed April 10, 2022.

United States Environmental Protection Agency (EPA). (2019). "Greenhouse Gas Emissions – Sources of Greenhouse Gas Emissions." https://www.epa.gov/ghgemissions/sources-greenhouse-gas-emissions. Accessed April 10, 2022.

Wagner, K. (2018). "City Noise Might Be Making You Sick." *The Atlantic*. www.theatlantic.com/technology/archive/2018/02/city-noise-might-be-making-you-sick/553385/. Accessed April 10, 2022.

World Health Organization: Regional Office for Europe. (2011). Burden of disease from environmental noise: Quantification of healthy life years lost in Europe. JRC European Commission. https://apps.who.int/iris/handle/10665/326424. Accessed April 10 2022

Chapter 4
Work and the City

Innovative digital technologies transformed our way of working. Nowadays, traditional office work can be completed remotely, providing employees greater control over working hours and redefining the work–life balance. The phenomena that started a while back experienced rapid growth during the COVID-19 lockdowns. Freed from long commutes and set schedules, many people enjoyed remote work, while businesses were able to eliminate the costs of renting and operating offices. It also contributed to environmental sustainability by lowering emission as a result of reduced commute by private vehicles. By designing effective home office spaces, for some, work from home (WFH) can be more productive. Alternative solutions such as coworking spaces can also provide much needed routine and quiet workspaces for "gig workers" outside the home but still close by. This chapter investigates the effects that digital means have on work, working from home, and design of such spaces.

4.1 The New Nature of Work

After decades of concentrating industry's location which gave rise to hubs like California's Silicon Valley, technological innovation threatens to render those regions and type of work obsolete. Some companies, like Twitter, have encouraged their employees to WFH indefinitely and begun downsizing their expensive office spaces to cut costs. The ability to WFH provides workers with the freedom to work from anywhere but will have untold impacts on regions like Silicon Valley. One can argue that nowadays the location of work can be in any place with access to Wi-Fi be it a café or a public outdoor workstation set by a city for use by all (Figs. 4.1 and 4.2).

Additionally, the emergence of the urban "gig economy" through applications like Uber or Postmates has changed the way people find employment in cities. Now

A. Friedman, *The Sustainable Digital City*,
https://doi.org/10.1007/978-3-031-25488-8_4

Fig. 4.1 Nowadays, a location of work can be any place with access to the Internet such as a café

casual or part-time employment can be picked up easily through app-based employment in cities. With a large push for WFH, especially in the wake of the COVID-19 pandemic, and increased opportunity for casual employment, the urban economy will move in a digital direction which may in turn create a more accessible opportunity for many. The pandemic has created a widespread need for suitable workspaces at home. The design and layout of a home office can significantly improve the well-being and effectiveness of a person's work. By accommodating one's specific needs and creating a healthy indoor environment, home offices can be comfortable and productive. Consequentially, the creation of coworking spaces was a response to the need for spaces outside the home for freelancers and creatives to work and foster social connections. Looking at the coworking model, one can imagine alternative work arrangements that address the needs of individuals while adapting to challenges in the digital economy and as result its effect on urban design.

As noted in previous chapters and will be outlined below, cities have become highly digital. Alert Ready, for example, is the emergency alert system used by Canadian authorities to notify the public about fires, natural disasters, missing children, and civil emergencies (AlertReady.ca., n.d.). After the alert, citizens can be shown the optimal route to their destination, coordinating transfers between transit modals and making their voyage more convenient (Dutzik et al., 2013). Creating, installing, maintaining, and updating technology in cities require workers, and in turn these technologies transform other professions.

Fig. 4.2 An outdoor public workstation in Montreal, Canada, offers free Wi-Fi access

4.1.1 The History of Working from Home

The contemporary nature of work looks remarkably different than it did in the late twentieth century. While historically people have worked from home, the nature of that work was different than it is today (Fig. 4.3). Until recently, large portions of a family's work took place at or near home, growing food for subsistence or working in small-scale businesses close to home and even in them. Following the technological advances of the industrial revolution, men, women, and children began working in factories. Working conditions were horrific, and many people became ill or died as a result. Women and men fought hard for fair wages and safe working conditions in strikes such as the 1888 Matchgirls' Strike against the Bryant & May match factory in London (Raw, 2009). This event inspired dock workers in London to organize against the harsh working conditions in the Dock Strike of 1889. These strikes sparked a movement to change working conditions; unions were formed to advocate for workers, protecting their health and ensuring fair wages.

As western economies transitioned from manufacturing to service economies, factories closed, changing employment opportunities. Also referred to by Shek et al. (2015) as a *knowledge economy*, a service economy is based on knowledge-intensive industries and services in economic production, well-educated workers in the occupational market, and innovative firms in business. Service industries are commonly made of enterprises that provide deliverables, which are intangible or immaterial. This economic transition led to the development of office buildings primarily

Fig. 4.3 Throughout history on streets like this one in Saint-Valery-sur-Somme in France, people resided in or above their place of work

located in city centers, creating the cityscapes we recognize today. Previously, many who could left urban areas to escape pollution and poor living conditions, moving to suburbs which provided open spaces, clean air, and an escape from the city. This attraction led to the urban sprawl on many cities taking up more land by growing out instead of up. This has forced people to live further from work increasing the length of daily commutes and resulting in a greater dependence on fossil fuels. In Canada, for example, the average citizen spends at least 1 h daily commutes to and from work (Statistics Canada, 2017). In addition to the environmental costs of contemporary urban planning, there are extensive social, physical, and psychological costs. Dependency on privately owned vehicles has normalized sedentary lifestyles. Coinciding with jobs becoming increasingly stationary, long commutes between the home, work, and essential services have resulted in limited physical activity increasing health problems, further burdening the healthcare systems.

Businesses began to recognize the strain that the workplace environment had on workers and explored alternative options. In the 1970s, IBM began experimenting with their *non-territorial offices*. The theory was that without permanent workstations such as cubicles and offices, workers would be free to find spaces within the office where they would be the most productive and collaborate with coworkers (Zhang, 2016). In 1973, the term *telecommuting* was introduced by Jack Nilles, an American researcher who developed the idea while stuck in traffic. Nilles theorized that eventually daily commutes could be eliminated altogether by allowing employees to WFH. By conducting research at an insurance company in Los Angeles, he found that it would be more cost-effective for businesses, increase productivity, and decrease environmental pollution (Zhang, 2016). However, he also found that a major setback of telecommuting was the social function that work plays in many people's lives. For many people, work is a major point of networking and social contact, so removing oneself to be isolated in at home was not appealing to many workers.

In the article *The Dark Side of Virtual Office and Job Satisfaction* (Zhang, 2016), Zhang studied employees' perception of communication, social relations, achievement recognition, and work–life balance in virtual office settings. Zhang found that there was a decrease in information richness, instances of a lack of attentiveness, and an increase in misinterpretations when working from home. The study demonstrates that while there are many benefits to working from home, such as increased personal time, increased ability to balance family responsibilities, and increased flexibility, there are some setbacks, and businesses should consider ways to mitigate these challenges (Fig. 4.4).

Since the 1970s, the proportion of people working from home has increased sharply. It has been estimated that in the United States and the European Union, approximately 40 percent of jobs can be completed from home (Gottlieb et al., 2021). Since March 2020, there has been a massive global shift toward WFH because of the spread of COVID-19. Many governments implemented physical distancing measures, requiring anybody who could to WFH in an attempt to limit the effects of the pandemic. A report by Statistics Canada (2021) recently released that in April of 2020 about 40 percent of Canadian employees worked from home—which supports Gottlieb's research. Comparatively four percent of Canadian employees worked from home in 2016. By August 2021, however, the percentage of Canadians working from home dropped to 23 percent, indicating that there was movement either by employees or employers to return to the office. While technologies such as Zoom, Microsoft Teams, and Cisco Webex exist to facilitate remote work, many businesses were not prepared to transition to strictly remote work as quickly as they had to. After a year of remote working, many businesses have adapted and created systems within the workplace to continue on with business.

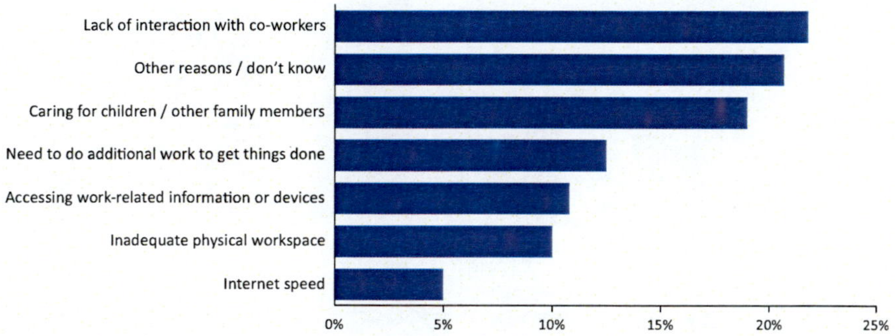

Fig. 4.4 Main barrier to productivity among teleworkers who now accomplish less work per hour

This sudden and rapid global shift to working from home leads to several questions about the future: will businesses return to working in offices or remain online? Will more employees have a choice to WFH? How will working from home impact the health and well-being of citizens? How will family structures be impacted by one or both parents working from home? How will networking and professional advancement change in an age with increasingly less in-person contact? It is important for employers to consider these and other questions and create healthy working conditions for employees (Fig. 4.5). One important factor in this discussion is the subjectivity and personal nature of this matter. While some workers might welcome and enjoy working from home, others might not, and different options should be available.

Other factors to consider when contemplating WFH is the toll that commuting takes on the environment. WFH or nearby coworking space help employees avoid hours long commutes to and from work. This reduces traffic on the streets, idling, and many unnecessary greenhouse gas (GHG) emissions. Lockdown measures that had many people working from home following the emergence of COVID-19 were clear evidence of this fact. Various studies such as that performed by Aunan et al. (2020) found that levels of nitrogen dioxide (NO_2) and fine particulate matter ($PM_{2.5}$) dropped significantly. In lockdown regions, on average, NO_2 concentrations were 60 percent lower, and $PM_{2.5}$ levels were 31 percent lower than expected (Aunan et al., 2020). While lockdown level mobility or lack thereof cannot be maintained, a return to pre-pandemic transportation habits must be reconsidered taking this information into account.

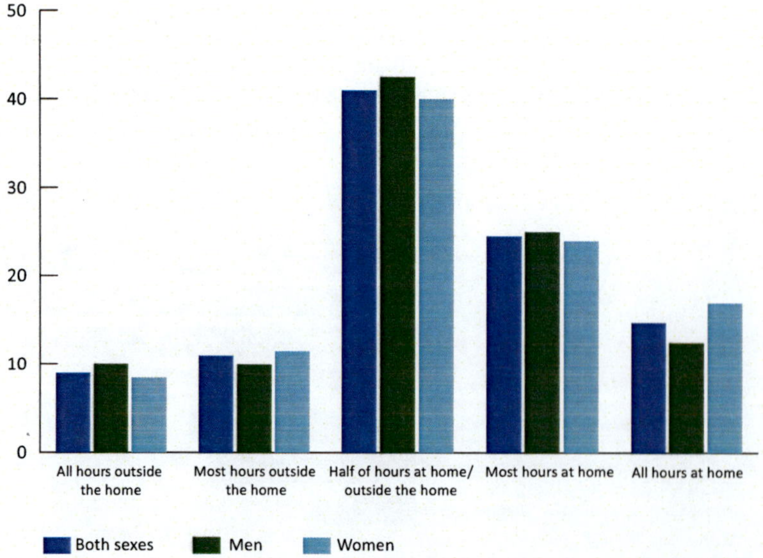

Fig. 4.5 Preference of new teleworkers to WFH once the COVID-19 pandemic subsides

It should be noted that while information technology has enabled many more professions to operate remotely, many other jobs cannot be done at home. Service jobs, healthcare, self-employment, agriculture, construction, and other essential services cannot be done remotely. For the purposes of this book, we will focus on industries that can be done at home, exploring the technologies and design strategies that contribute to productive and successful at-home work environments.

4.1.2 *Connectivity*

As noted above, innovation of communication technologies has enabled people to rapidly connect from anywhere with virtually anyone at any time. This freedom of communication has changed our daily interactions, contributing to an increasingly globalized society and instantaneous access to information. To reiterate from Chap. 1, around 60 percent of the world's population uses the Internet, and over 5 billion people use mobile phones (Abdulla & Esmaeel, 2019; World Bank Group, 2020). This connectivity also has revolutionized the nature of work, created new jobs, and changed the structure and places of traditional employment. With sufficient Internet and mobile connectivity, many employers no longer require workers to be in a physical office. Businesses around the world are no longer constricted by geographical boundaries. Employees can work wherever they are in the world for companies whose offices might be thousands of kilometers away (Newman, 2020). Many major technology companies such as Google, Facebook, and Twitter have made extended or permanent arrangements for employees to continue working from home, eliminating the need for workers to even live in the same country as their job (Newman, 2020). A controlled study by Bloom et al. (2015) entitled *Does working from home work? Evidence from a Chinese experiment* found that working from home led to a 13 percent increase in productivity and 50 percent less turnover in staff working from home; however, there was a decrease in rates of promotion among those working from home. Following the experiment, the company offered all employees the option to WFH which many workers chose. According to Bloom et al. (2015), the firm improved total factor productivity by between 20 percent and 30 percent and saved about US$2000 a year per employee who worked from home.

New technologies that enable information sharing and real-time collaboration have determined which industries can function with their employees working from home. Email has changed the way in which employers communicate with workers. Workers can be located anywhere, provided they have access to a digital device and the Internet. Applications such as Google Docs allow users to share, view, and edit the same document at the same time while automatically saving and storing data. Such functions limit the incidents of losing files, working on different versions of documents, and having to distribute hard copies of information.

This freedom to WFH has benefited many; however, there is a complicated social, personal, temporal, and physical transition that needs to be navigated by workers (Crosbie & Moore, 2004). When transitioning to working from home, a new work–life balance needs to be established in the interest of balancing occupational stress, personal time, and in some instances childcare. In their study, Crosbie and Moore found that many at-home workers enjoyed the increased flexibility in delineating their personal time and time to care for others from work. However, many people interviewed noted there are challenges such as loneliness and a lack of self-esteem (Crosbie & Moore, 2004). Colleague support networks, personal achievements, and socializing with coworkers were all greatly diminished when working from home. Many businesses presently aim to mitigate this through virtual happy hours and organized online social gatherings.

4.2 Contemporary Work from Home

When discussing the transition to WFH, it is important to consider type of work that one is doing and the space available to performing it in the worker's residences. To begin with, according to Dietsch (2008), a workspace should be separate from the living space. Unfortunately, many homes often do not have spaces that are dedicated to work. A home office is not simply a spare bedroom or nook in the corner of another room. Spaces like those can blur the lines between work and home life and create distractions, making it difficult to focus on work. If other members of the household are at home during the day, it is especially important to have a designated working space. Individuals that meet with clients need spaces that are clearly separated from the home to maintain professional boundaries and a large enough office to comfortably sit two or more people. Unfortunately, the home-building industry has been slow to adapt to these new demands. Just as there are a set of design principles that can make offices calming, comfortable, and focused environments, at-home offices are most successful when they follow similar guidelines.

4.2.1 Creating a Productive Space

The indoor air quality, thermal control, visual quality, and acoustic level of a space greatly impact the health and well-being of its users. In a space such as a home office, it is especially important that each of these factors is considered and adjusted to the user's preference. Sufficient daylight and a quiet space are the most conducive to productivity. While artificial light is needed for visibility, regulating one's workday by glare-free daylight can help set a work schedule and improve an individual's mood (Fig. 4.6). If noise from the living space is audible in the workspace, it can be distracting. Placing rugs on the floor, sound-absorbing panels on the walls, and adhesive strips along the bottom of the door can help decrease noise and create a calmer space. If the space is too warm or cool for comfort, adding a fan, space heater, humidifier, or dehumidifier can improve the thermal regulation of the space. Air quality is largely dependent on the systems used in the house, but when working

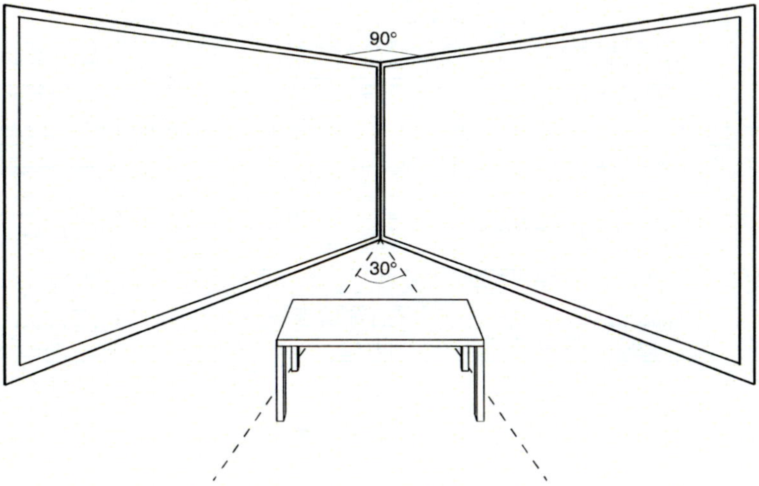

Fig. 4.6 A workspace should be in an illuminated but glare-free area

from home, it is especially important to ensure the air systems are clean and functioning properly. Having windows that open to let in fresh air is also important (Dietsch, 2008).

While there is the possibility to retrofit a space to accommodate a home office, it is cheaper and more efficient to design homes with the potential for having a home office. Without proper visual, spatial, and acoustical separation, the distractions of domestic life are unavoidable—no matter how focused the individual. Despite the decision to WFH, there still exists a desire to "go to work." Separating a home office from the main living space can be done by locating the office on a different level such as the attic, attaching a new unit to the side of the dwelling, or constructing a new smaller dwelling behind the existing home (Figs. 4.7 and 4.8). Adding a second entrance to the dwelling that leads to the office can be done in cases where clients will be visiting. Flexible partitions are a good option to subdivide existing spaces that are underused. It allows the homeowner to delineate a clear working space while allowing for future changes. Establishing a healthy work–life balance is widely cited as the key to a successful WFH arrangement. Balancing personal and work responsibilities varies depending on one's home life and job. Some might find a traditional 9 am to 5 pm schedule works for them, while others might prefer to do work in the evenings (Crosbie & Moore, 2004). Regardless, the physical space in one's home is important for productive work and enjoying personal time.

There are a series of design changes that can contribute to an effective home office such as storage shelves, electrical outlets, telephone and fax lines, sufficient table space, a comfortable chair, and task lighting. Depending on the nature of work, shelving to store reference materials is important to keep the space organized and comfortable (Fig. 4.9). Sufficient electrical outlets including adaptable ones will enable more variations for locating furniture in the workspace (Fig. 4.10). Telephone

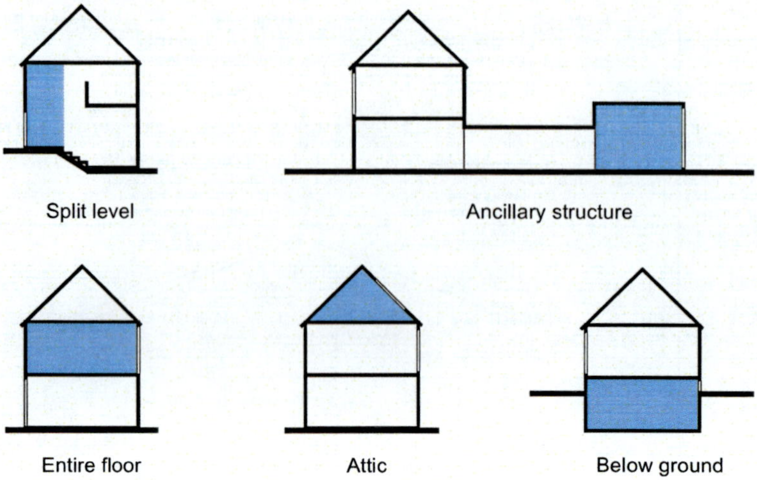

Fig. 4.7 Possible locations of a home office in a dwelling

Fig. 4.8 A home's main floor (bottom) or the attic (top) can be the locations of workspaces

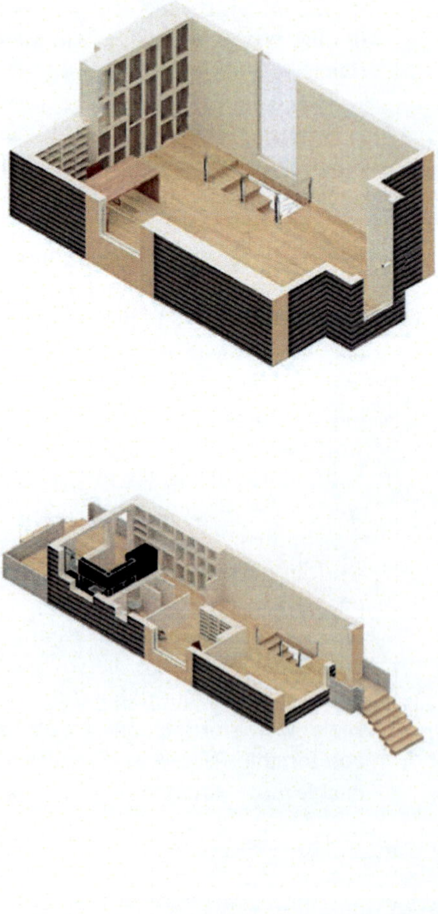

Fig. 4.9 Depending on the nature of work, shelving to store reference materials is important to keep the space organized and comfortable

Fig. 4.10 Sufficient electrical outlets including adaptable ones will enable more variations for locating furniture in the workspace

and fax lines, while potentially unnecessary for many professions, are still important to others and should be found in a workspace. Appropriate furniture such as a desk of a correct height and size will contribute to the comfort of a space as well as a chair that is ergonomically designed to prevent poor posture. Task lighting as well as ambient lighting is critical for a comfortable work environment; soft but bright overhead lighting contributes to productivity as does more direct, smaller light sources located close to spaces where they are needed such as a desk lamp or reading light.

Keeping the workspace organized is important to maintain a professional environment as well as be more productive. An at-home office is an opportunity to create a comfortable and personal work environment, tailored to the user's needs. Painting and decor choices are very personal and should be considered by the individual but should contribute to a bright, productive environment. Plants, art prints, and other decorations can be used to brighten a space and create an environment the individual enjoys being in. Video conference calls have become an almost daily task for many people working from home, and as a result it is important that the workspace is presentable and professional. Reliable equipment is also essential when working from home; a stable and reliable Internet connection and updated computer with the required programs are necessary for the majority of at-home work.

4.2.2 Adapting Professions and Education

Many professions have transitioned to function wholly or partly online. Though exacerbated by COVID-19, this transition started long before 2020. Teaching is a prime example of how technology has altered the traditional structure of education and role of professors. Open University, a research institute in the United Kingdom, has pioneered incorporating technology into higher education, making university degrees accessible to a wider portion of the population. Founded in 1969, Open University "extended higher education beyond a typical class of students by using media as a tool to transform both the production and transmission of knowledge" (Moreno, 2017). By securing prime time slots on television and radio, Open University provided courses to the working population, aiming to provide the opportunity to pursue a higher education to those who otherwise wouldn't have the time or access. Physical materials such as workbooks were mailed to residences, and students followed lectures and interviews that were scheduled between 3 pm and 6 pm (Moreno, 2017). Students could be located anywhere in the United Kingdom and complete courses on a wide range of topics. Open University, which is still operating today, spearheaded the distanced learning and adult education movement by using technologies available at the time.

Education has continued to adapt and incorporate technologies to improve teaching and make higher education accessible to a wider population. Prior to 2020, many universities were already offering a portion of their classes online, while other institutions such as Athabasca University in the Province of Alberta, Canada, operate through online distance education. In traditional classroom structures, technology is a powerful asset for professors and students. Lectures are posted online for reference throughout the semester, students are able to ask professors questions online through messaging platforms and email, and virtually every student uses a laptop to take notes and complete assignments. Even in primary and secondary school, technology has revolutionized teaching. Smart Boards and laptops in classes are valuable tools for teachers, and education technologies have been used to accommodate different learning styles and needs. In the wake of COVID-19, many universities have transitioned to temporarily offer all classes online. Some institutions have begun offering a select number of free online classes for anyone to take, further increasing accessibility to education and inspiring people to pursue new interests. More on the applications of technology in the education sector will be covered in Chap. 6.

Outside of formal institutions, other businesses and organizations offer online training on a wide variety of topics. The possibilities are nearly endless. Webinars ranging from software, coding, graphic design, web design, music production, creative and technical writing, photography, to social work and mental health advocacy, to name a few, are available to anyone. Platforms such as YouTube host free "how to" videos on millions of topics. For many people, creating online content and hosting webinars is a significant source of income, creating new jobs for themselves and others. Technology has revolutionized the job market, creating thousands of new knowledge-based jobs.

4.3 Working Alone Together

While the ability to WFH saved many businesses in 2020 throughout COVID-19 lockdowns and benefited a great deal of employees, many find it difficult and challenging compared to the traditional office setting. Collaboration and productivity can be improved by being in the same physical space as another person, and many people enjoy the social aspect of work forging new friendships with coworkers. As discussed above, there is variability across professions as well. Call centers might operate better when employees WFH, whereas creative professions with extensive collaboration between employees are more challenging to do remotely. Additionally, nonstandard forms of employment have become commonplace within a highly individualized labor market in which urban professionals work as a casualized, project-based, and freelance workforce (Gandini, 2015). Often referred to as the "gig economy," it encompasses jobs such as freelancers, independent contractors, delivery drivers, musicians, truck drivers, tradespeople, and many other professions. The rise of gig work began following the 2008–2009 economic collapse and has become an increasingly popular form of employment; in 2018, 150 million people in North America and Western Europe worked as independent contractors (Petriglieri et al., 2018). The gig economy can be divided into two distinct categories, knowledge-based and service-based gigs.

Knowledge-based work is driven by technology, software platforms that enable the sharing economy (Frazer, 2019). This work, previously done by people working full time for a larger company, is now done by individuals who are hired on short-term contracts. This structure transfers an enormous amount of responsibility from the business to the individual, but many have noted that the personal benefits outweigh the stress. The Harvard Business Review found that most gig workers surveyed maintained valuable connections with "place, routines, purpose, and people'" to succeed in their roles (Petriglieri et al., 2018). This further supports the need for suitable home offices that accommodate the individual's needs.

There are alternative work structures between the traditional office setting and employees working from home. Some businesses have allowed employees to WFH for a set portion of the week such as 2 days at home, 3 days in the office, while other businesses simply let employees choose where they would like to work, at home or in the office. What happens, in these instances, when a person either does not have space available to them at home to work or does not like working at home? Or what if a business is completely remote but requires some in-person meetings? In response to these and other issues, coworking spaces have been created in many urban areas (Fig. 4.11). The first official coworking space opened in 2005 in San Francisco as a hybrid or "third way" to work—not an office, not remote work (Gandini, 2015). Founded on principles of collaboration, openness, community, and sustainability, coworking spaces were not designed to be a replacement for offices; rather, they were intended to foster connection between creatives and an open-sourced community approach to work (Gandini, 2015). Coworking spaces can therefore be regarded as a new form of urban social infrastructure enabling contacts and collaborations between people, ideas, and connecting places (Merkel, 2015). While some more

Fig. 4.11 Space constrain and a desire to work in the company of others led to the creation of coworking spaces like this one in Istanbul, Turkey

closely resemble a cafe that is specifically tailored to a quiet work environment, others appear more similar to modern office spaces with conference rooms and private rooms available to rent daily. These spaces have been designed for modern work hosting fast Internet, comfortable and varied seating options, sufficient electrical outlets, and televisions for presentations. Many coworking spaces have separate areas for individual and group work. Coworking spaces can be used by freelancers, entrepreneurs, artists, and remote workers. They provide flexibility, community, an opportunity to leave the house, and at times networking. Coworking spaces foster personal connections and routines that improve productivity and success at work. As mentioned previously, however, with regard to the gig economy, coworking spaces also transfer responsibility from the business to the individual. Instead of the business being responsible for owning or renting and operating an office space that provides the necessary equipment for its employees, workers must rent their own space, pay for a reliable Internet connection, and own their own equipment.

One alternative is for a collection of businesses that primarily function remotely to maintain a smaller office, with spaces available for several companies in the building. The business would be responsible for renting the space and providing the required equipment but with a smaller capacity. By combining coworking practices and traditional office structures, the financial burden would not be the responsibility of workers but would still decrease operation costs. By combining multiple businesses in the same space, operation costs would be decreased further.

4.4 Reflections

Rapid technological development has dramatically changed the function of cities and employment opportunities. Technology has revolutionized the way many jobs are done and has created plenty of new ones. For example, just the development and maintenance of technologies themselves has created new jobs and facilitated the creation of other positions. App development such as Twitter, Facebook, Uber, and Airbnb has contributed to the expanding gig economy. Knowledge and creative intensive industries have become a significant employment sector and in turn have changed the traditional work structure. As long commutes to distant offices become less requisite and associated fees become more expensive for both employees and employers, many companies are encouraging part-time or full-time WFH. Following the COVID-19 pandemic, many businesses have made a permanent transition to remote work. Working from home provides employees with greater control over their work–life balance, decreases the cost of commuting, and improves productivity. While decreased social interaction is a significant setback, many remote workers have noted the benefit of increased personal freedoms.

Well-designed home offices are a contributing factor to successfully working from home. Creating a healthy indoor environment conducive to concentration that is tailored to the user's job requirements will contribute to productive work. Clearly separating the living space from the workspace is a vital component of a home office. Installing appropriate lighting, heat controls, and sound buffers will create a comfortable space; sufficient storage space, an ergonomic chair, desk, and reliable equipment (such as a telephone and computer) will improve the function of the space. While the specifics in design will vary by profession and personal preference, the previously explored standards of privacy, comfort, and organization apply to any home office.

Technology has vastly changed how we learn as well as how and where work can be done. Coworking spaces have evolved as an alternative to traditional office spaces and home offices. By creating spaces for entrepreneurs and freelancers, coworking spaces have become hubs of innovation and social connection while preserving individual agency. By creating spaces for both self-employed and remote workers to use, social connections are fostered, and community is supported. Ideally innovative solutions can be created through collaboration, and the wider community will benefit. Technology has, and will continue to have, a monumental impact on our lives, the jobs we have, and the cities we live in.

Questions for a Follow-Up Discussion

1. How did the nature and place of work change since the COVID-19 pandemic?
2. What is the role of technology in expanding telecommuting? In what way?
3. How can proper architectural design make work and home life mutually exist in dwellings?
4. Will working "alone in commune" become a permanent practice?
5. What are the advantages and disadvantages of online meetings? Are they about to replace in-person meetings in the future?

References

Abdulla, M. A., & Esmaeel, A. M. (2019). *Providing information through smart platforms: An applied study on academic libraries in Saudi universities* (pp. 1–24). Journal of Education.

AlertReady.ca. (n.d.). Alert ready: Emergency alert system. https://www.alertready.ca/. Accessed April 13, 2022.

Aunan, K., Chowdry, S., Lelieveld, J., & Venter, ZS. (2020). "COVID-19 lockdowns cause global air pollution declines". Earth, Atmosphere, and Planetary Sciences. https://doi.org/10.1073/pnas.2006853117 Accessed April 14, 2022.

Bloom, N., Liang, J., Roberts, J., & Ying, Z. J. (2015). Does working from home work?. Evidence from a Chinese experiment. *The Quarterly Journal of Economics, 130*(1), 165–218. https://www.gsb.stanford.edu/faculty-research/publications/does-working-home-work-evidence-chinese-experiment. AccessedApril 13, 2022, 130, 165

Crosbie, T., & Moore, J. (2004). Work-life balance and working from home. *Social Policy and Society, 3*(3), 223.

Dietsch, D. (2008). "Live/work: Working at home, living at work". *Abrams.*

Dutzik, T., Madsen, T., & Baxandall, P. (2013). "A new way to go: The transportation apps and vehicle-sharing tools that are giving Americans the freedom to drive less". *US PIRG education fund, frontier group.* https://uspirg.org/sites/pirg/files/reports/A%20New%20Way%20to%20Go%20vUS1_1.pdf. Accessed April 13, 2022.

Frazer, J. (2019). "How the gig economy is reshaping careers for the next generation". *Forbes.* https://www.forbes.com/sites/johnfrazer1/2019/02/15/how-the-gig-economy-is-reshaping-careers-for-the-next-generation/?sh=7738282549a. Accessed April 13, 2022.

Gandini, A. (2015). The rise of coworking spaces: A literature review. *Ephemera, 15*(1), 193–205.

Gottlieb, C., Grobovšek, J., Poschke, M., & Saltiel, F. (2021). Working from home in developing countries. *European Economic Review, 103679*, 103679.

Merkel, J. (2015). Coworking in the city. *Ephemera, 15*(2), 121–139.

Moreno, J. (2017). The university is now on air: Broadcasting modern architecture[exhibition]. Exhibited at the Canadian Center for Architecture November 15 2017 - April 1 2018.

Newman D. (2020). "Distributed companies, work from home, and the technology enabling the change". *Forbes.* https://www.forbes.com/sites/danielnewman/2020/10/22/distributed-companies-work-from-home-and-the-technology-enabling-the-change/?sh=16b93c0033bf. Accessed April 13, 2022.

Petriglieri, G., Ashford, S. J., & Wrzesniewski, A. (2018). "Thriving in the gig economy". *Harvard Business Review.* https://hbr.org/2018/03/thriving-in-the-gig-economy. Accessed April 13, 2022.

Raw, L. (2009). Striking a light: The Bryant and may matchwomen and their place in labour history. *A&C Black.*

Shek, D. T., Chung, P. P., & Leung, H. (2015). Manufacturing economy vs. service economy: Implications for service leadership. *International Journal on Disability and Human Development, 14*(3), 205–215.

Statistics Canada. (2017). "Journey to work: Key results from the 2016 census." https://www150.statcan.gc.ca/n1/daily-quotidien/171129/dq171129c-eng.htm. Accessed April 13, 2022.

Statistics Canada. (2021). "Working From Home in Canada: What have we learned so far?". https://www150.statcan.gc.ca/n1/pub/36-28-0001/2021010/article/00001-eng.htm. Accessed April 13, 2022.

World Bank Group. (2020). "Individuals using the internet (% of population)". https://data.worldbank.org/indicator/IT.NET.USER.ZS. Accessed April 13, 2022.

Zhang, J. (2016). The dark side of virtual office and job satisfaction. *International Journal Of Business And Management, 11*(2), 40.

Chapter 5
Efficient Energy Distribution and Waste Management

By 2050, it is expected that global energy consumption will increase by a third and waste production will nearly double. Yet, the implementation of sustainable practices into the energy production and waste management sectors has the potential to significantly reduce greenhouse gas (GHG) emissions and mitigate climate change. This chapter reviews key concepts of the electricity grid and how smart technology is changing it. Included is the decentralization of energy through techniques such as district heating (DH) and microgrids. Following, waste management is addressed first through techniques of waste elimination via digital and other means and second by exploring digitally enhanced waste collection methods used for unavoidable waste.

5.1 The Need for Alternative Power and Conservation

Energy generation and consumption is a notable topic for good reason. On a global scale, the energy sector presently contributes around 72% of anthropogenic GHG emissions (Center for Climate and Energy Solutions, n.d.). The highest polluting industries within the energy sector are electricity, heating, transportation, and construction which contribute 31%, 15%, and 12.4% of global GHG emissions, respectively. The causes of these high GHG emission levels lie in the fact that 80% of the world's energy is supplied by fossil fuels such as coal, oil, and natural gasses (IEA, n.d.). Additionally, demand for energy is growing globally with an expectation that energy consumption will increase by a third between 2017 and 2035 (Bollen, 2017). There is an evident incompatibility between the finite fossil fuel sources for energy and a seemingly infinite demand—not to mention the environmental implications of failure to reduce GHG emissions within the coming years. The need for alternative energy sources, which presently meet just 20% of the world's energy needs, has never been clearer and simultaneously has never been so supported. Technology,

A. Friedman, *The Sustainable Digital City*,
https://doi.org/10.1007/978-3-031-25488-8_5

research, and funding throughout recent decades have made the switch away from fossil fuels toward renewable energy sources (RESs) more feasible now than ever. Renewable energy which is derived from natural processes that are replenished at a rate that is equal to or faster than the rate at which they are consumed (Canada, 2017) is complementary to the sustainable agenda outlined in this book. RESs allow for the present population to meet their energy needs, without infringing on the ability of future generations to do the same.

Technological advances made in recent decades have popularized the availability of RESs despite them having been present for hundreds of years prior. Water wheels, for example, predicate modern-day hydro power plants by converting kinetic energy from moving water into usable mechanical energy (Project Solar UK, 2018). Likewise, windmills have been used since the sixteenth century for the use of wind energy. Today, hydropower and wind energy are valuable RESs and exist among others such as solar, geothermal, biomass, and hydrogen. These energy sources have been optimized and automated through the use of digital technologies and are now viable solutions to the impending energy crisis. Hydro- and wind power have both advanced significantly since their introduction (Fig. 5.1). Hydroelectric dams and power plants continue to transform kinetic power to usable energy in an increasingly efficient manner, and wind power has evolved to capture offshore wind energy, utilizing ocean winds to generate electricity on either fixed or floating turbines (Canada, 2017; United States, 2021b). Solar power harnesses the energy of the sun, the most powerful source of energy available to us on Earth. It is also free and available in abundance in certain climates. The invention of solar photovoltaic (PV)

Fig. 5.1 Produced in locations like this one near Glasgow, Scotland, wind power has advanced significantly since its introduction

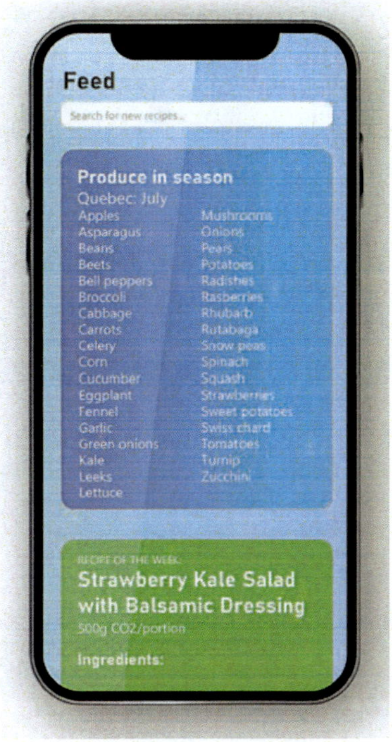

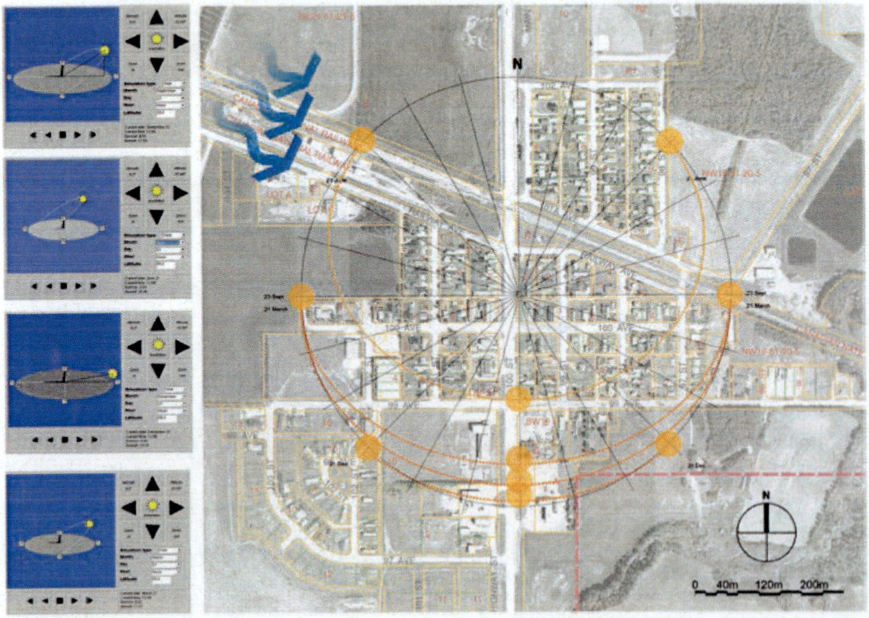

devices means sunlight can be converted directly into electricity without creating a single pollutant (United States, 2021b). Other technologies, such as concentrated solar thermal systems (CSP), convert energy from the sun less directly than PV devices. Solar energy requires solar panels to be installed on building roofs or solar farms and as of recently can be stored in various forms.

Geothermal energy comes from heat produced within the Earth's core which contains temperatures of up to 5982 °C (10,800 °F) (United States, 2021b). Geothermal power plants pump steam up from a few miles underground which is then used to spin turbines that produce energy (Fig. 5.2). In a similar fashion, geothermal heat pumps can be used to push water or other refrigerants through pipes below the Earth's surface where the temperature is a constant 10 °C to 15 °C (50 °F to 60 °F) (EPA, 2017). These systems can be used for heating in the winter or cooling in the summer (EPA, 2017). Biomass, such as wood, agricultural crops, animal

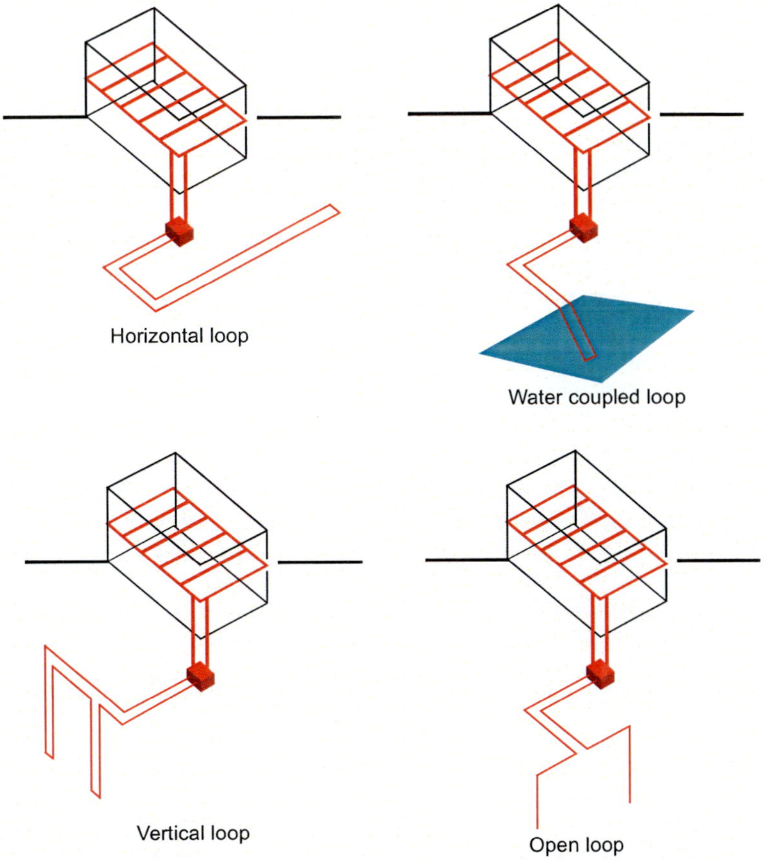

Fig. 5.2 Geothermal power plants pump steam up from an underground source which is then used to spin turbines that produce energy

manure, and human sewage, and other biogenic materials, such as paper, cotton, and food, among other waste products, contain sunlight in the form of chemical energy (Canada, 2017; United States, 2021b). With specific equipment and technology, these materials can be converted to energy through direct combustion—burning (which produces GHGs)—or one of thermochemical, chemical, or biological conversions to produce solid, liquid, or gaseous fuels such as biogas (United States, 2021b). Unlike the abovementioned RESs that are dependent on climate and geographic features, different types of biomasses are more widely available in abundance irrespective of physical location. These RESs can also be used to power other clean energy sources such as hydrogen power. Hydrogen power is created through a process called electrolysis in which water is split into its oxygen and hydrogen atoms (Enbridge Gas, n.d.). The hydrogen can then be used to power hydrogen vehicles or as an alternative source of energy. It can also be used for hydrogen blending, a term used to describe when hydrogen is injected into a natural gas supply to lower its carbon footprint (Enbridge Gas, n.d.). This is a useful technique during the transition toward renewable energy.

As evident, RESs are nuanced in their availability with different geographic regions supporting different sources. In Iceland, for example, geothermal energy accounts for 66% of energy use in the country (NEA, n.d.). Their proximity to a continental rift with high volcanic activity and therefore warmer surfaces below the Earth's crust makes geothermal energy easily accessible with the right tools. In a similar manner, a desert country such as Egypt might opt for solar power over hydropower. While past issues such as energy storage are becoming increasingly irrelevant, a sensitivity to the cost of transporting energy must be established to promote a sustainable industry.

A monumental moment for RESs occurred in 2020 when the International Renewable Energy Agency (IRENA) published their annual report stating that the cost of using renewable energy has for the first time undercut the cost of fossil fuels (IRENA, 2021). While the use of RESs themselves has always been cost-effective, the manufacturing, installation, and upkeep of infrastructure required significant costs and have only recently been incorporated into economies of scale large enough to significantly reduce costs. The development of most cost-conscious technologies has also aided in the mitigation of prices (IRENA, 2021). Not only do RESs promote environmental and now economic sustainability, but many countries are also tied as signatories to uphold international agreements made to support the transition to clean energy. At the 2021 United Nations Climate Change Conference otherwise known as COP26, countries such as Canada, the United States, Costa Rica, Finland, and others to a total of more than 20 committed to ending public financing of fossil fuel projects abroad by the end of 2022 (Abnett & Jessop, 2021). Similarly, the historic Paris Agreement, a legally binding international treaty signed by 196 countries at COP21, committed participating nations to limit global warming to at least 2 °C (3.6 °F) below pre-industrial levels (UNFCCC, n.d.). These international agreements hold countries accountable to each other and their citizens, provide support, and monitor the progress on the way to achieving set goals. These conferences, treaties, and subsequent monitoring are all assisted by digital means and contribute globally to the reduction of GHG emissions. As will be seen in the coming chapters, there are many different energy conservation techniques afforded by technology that complement the shift to renewable energy.

5.2 Digital Contributions to Managing the Energy Grid

Much of everyday life in the twenty-first century is dependent on easy, quick, and reliable access to limitless amounts of electricity. Charging phones, computers, and electric vehicles, using any electric household appliances such as toasters or refrigerators, and turning on the lights at home or in the office all require electricity at a moment's notice. All of this is provided by a complex infrastructure known as the electricity grid or the energy grid (Fig. 5.3). The modern electricity grid consists of four main components: the generation of energy, followed by the transmission, distribution, and consumption of electricity (Student Energy, n.d.). To start, energy is generated at large power plants, often far from consumers. This energy can be generated by any of the abovementioned RESs or by one of coal, nuclear power, or natural gas and is subsequently converted to electricity. Post-generation, substations also known as transformers "step up" electricity voltage for transmission because electricity losses are minimized at higher voltages (Student Energy, n.d.). Electricity is then transmitted at a high voltage along power lines to a desired consumption site. Upon arrival closer to the consumption site, substations "step down" the voltage so that the electricity can be distributed to its final site which can be industrial, commercial, or residential. The modern grid has come a long way in optimizing electricity distribution. Just over 100 years ago, the electricity grid did not exist, and individual low-voltage electricity generators operated in isolation serving local customers through short distribution lines (United States, 2021a). As demand increased over time, utilities began to connect their transmission systems so that numerous power generation plants could transmit electricity further and/or where it was needed most. This union of systems allowed for more cost-effective and reliant delivery of electricity especially at peak demand times and has evolved into the electricity grid we know today.

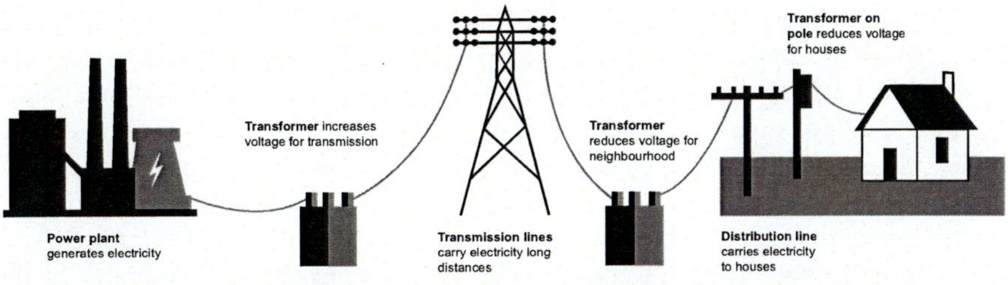

Fig. 5.3 The modern electricity grid consists of four main components: the generation of energy, followed by the transmission, distribution, and consumption of electricity

Though leaps and bounds have been made in terms of efficiency of the present-day electricity grid, there are still varying arising issues. To begin, in many regions, much of the infrastructure supporting the grid is outdated. In the United States, for example, more than 70% of power lines are more than 25 years old (Shively, n.d.). The wear and tear sustained by these electrical apparatuses heightens all the other impending issues, such as electricity loss. Between 5% and 30% of all generated energy is lost within the grid (Clean Energy Grid, n.d.). Most of these losses occur as heat when energy is converted into electricity, and a constant 5% is also lost as heat as electricity is transmitted and distributed to consumption sites. These losses are increasingly felt as energy demand patterns change and increase. The US Energy Information Administration (EIA) predicts that there will be nearly a 50% increase in world energy usage from 2010 to 2050 (Kahan, 2019). Much of this growth will occur in the industrial sector, in regions where strong economic growth is driving demand, particularly in Asia (Kahan, 2019). There will also be a larger electricity demand from the transportation sector worldwide as vehicles shift toward being electrically powered. Similar to changing usage, changing energy sources (from fossil fuels to renewables) is also not a small feat. The unpredictability of solar and wind energy means grids must increase system flexibility through energy storage capabilities and integrating alternative energy sources. The vitality of the electricity grid means these problems must be addressed, a complex task that is largely completed through digital means.

The smart grid is the digitization of the present electricity grid and one of many solutions to managing growing energy demands. It uses information communication technology (ICT) to transform the present grid into an intelligent network that will improve the reliability and efficiency of services, enhance markets, and reduce costs (IRENA, 2015). One of the most important capabilities of smart grids is that of two-way communication of information detected via monitors or customers to central operators. Information communicated can consist of infrastructure health, energy usage, and more. Smart meters, for example, will capture electricity consumption on an incremental basis (often hourly) (IESO, n.d.). This will allow for both users and suppliers to adjust their demand responses accordingly. Users might choose to opt for smart, electricity-saving appliances. Suppliers on the other hand can reduce the stress of peak demand times by equipping themselves with appropriate energy storage capabilities. Smart grids also feature advanced inverters, equipment that is able to convert energy generated by solar panels to energy usable in an electrical grid which will help facilitate the integration of RESs (Energy.gov, n.d.). Another capability of the smart grid is its disaster response ability. For example, as the frequency and severity of storms increase, so will power outages. In these instances, smart grids will be able to detect and isolate the outages to either stop them from occurring completely or at least minimize the impact by automatically rerouting when failures occur (SmartGrid, n.d.). For a century, the grid system has been the main source of energy for many countries. As energy sources change however, so will the grid. Decentralized control of energy, also facilitated by the digital age, is another electricity management technique that can be used to complement the smart grid and will be discussed in the next section.

5.3 Digital Means in Decentralized Heating

The above-described electricity grid is characterized by a unidirectional flow of energy that is generated at a significant distance from the use site. Decentralized energy (DE) on the other hand, as defined by Decentralised Energy Canada, is the heating, cooling, or electrical energy that is produced, managed, and/or stored at the point of consumption (DEC, n.d.). DE refers to several technologies that lessen electricity demand on the main grid. The implications on the grid of DE are distributed generation and storage of electricity, energy efficiency, and demand response. DE systems reduce the demand on the main grid especially during peak hours and offer more autonomous energy sources while maintaining grid connection. Solar panels, district energy, and microgrids are all different examples of DE. They all generate energy in situ and, as elements of a smart grid, can be equipped with ICT and other smart technologies to maximize efficiency.

Solar panels installed as described above make use of free energy received from the sun. The installation of solar panels harnesses immense amounts of energy. In Canada, for example, there is enough insolation that solar panels installed on the roofs of residential buildings could meet half of the country's entire residential electricity demand (Canada, 2017). Panels installed on building roofs relieve the grid of the same amount of energy provided by them resulting in cost savings, increased property value, and clean energy (United States, n.d.). Not only is solar energy inherently cheaper than that purchased from the grid, but there are also various other monetary incentives to installing solar panels (Fig. 5.4). To start, many energy providers will buy excess energy from buildings with solar panels connected to the grid. With the installation of an extra inverter, after having met household electricity needs, extra energy will be off-loaded to the grid, and a credit will be given to the solar panel owner (Vivint Solar, n.d.). This is a process known as net energy metering (NEM). Similarly, there are many rebates offered through government programs. In the United States, for example, there is a federal investment tax credit that will cover up to 30% of the installation costs of solar panels.

Another form of DE is district energy, otherwise known as DH. It is a central energy source that generates heat for an entire neighborhood or region (Fig. 5.5). These systems can source energy via fossil fuels but are also compatible with solar-, wind-, geothermal-, or natural gas-generated power, to name a few renewable sources. These energy sources are used to heat or cool water and distribute it through pipes throughout communities, heating and cooling buildings and providing hot water as needed (Friedman, 2022). DH technology is unique in its ability to use both the heat and electricity generated from power sources, as opposed to the grid which captures just the electricity, making it a viable environmentally and economically sustainable option. It also eliminates the need for each building to have individual boilers, chillers, heat pumps, and cooling towers (Creative Energy, n.d.). DH can be installed at many scales such as college campuses, hospitals, and neighborhoods.

Helsinki, Finland, is a good example of a successful DH system. Being the northernmost and coldest capital city on the European continent, Helsinki has substantial

Fig. 5.4 Solar water heaters (top left), condescend tubs (top right), solar panels (bottom right), and small-scale wind turbines (bottom left) installed on buildings

heat demands throughout its winters. DH is able to meet 90% of the heating needs of the city through 14 different power and heating plants. One of them, the Katri Vala heating and cooling plant in use since 2006, is the world's largest. It produces heating and cooling from sea water and from waste heat of purified sewage water (HELEN, n.d.). From its DH systems, Helsinki saves the same amount of energy that would be consumed in the heating of 270,000 single family homes annually. The city's air quality is also one of the cleanest in the world in relation to other capital cities.

DH is a more efficient method of providing heat and hot water. The required infrastructure for DH, however, is substantial and must be strategically planned for successful integration and future expansion. Transitioning existing infrastructure to DH is a significant financial commitment for both individual homeowners and the neighborhood. Homeowners and businesses must retrofit their heating systems and facilities, and piping must be built (Friedman, 2022). In the European Union, approximately 9% of heat demand is met via DH; however, 40% of this is powered by gas, and 29% is powered by coal (Lichtenwoehrer et al., 2019). The conversion of these systems to being powered by RESs is critical to environmental success and meeting international agreements to limit the impacts of climate change.

Another example of DE is a microgrid. This is a self-sufficient energy system that serves a discrete geographic footprint such as a college campus, hospital

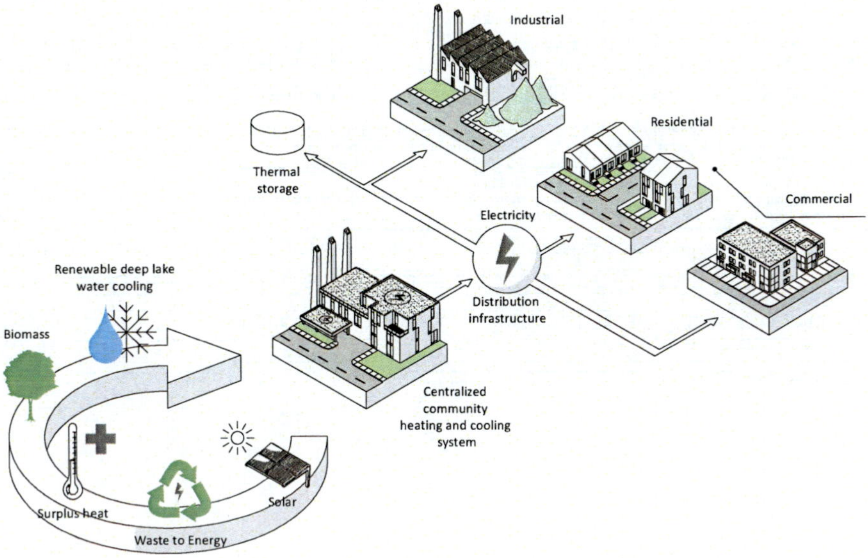

Fig. 5.5 DH is a central energy source that generates and supplies heat for an entire neighborhood or district

complex, business center, or neighborhood (Wood, 2020). Microgrids often use various power sources, have energy storage capabilities, and incorporate other forms of DE such as solar power and DH as described above (Fig. 5.6). They are energy self-sufficient and are able to fully disconnect from a central grid. Like main energy grids, microgrids use ICT to maximize efficiency, an example being switching back and forth between power sources for optimal electricity reliability, cost, and cleanliness of energy. Most microgrids remain connected to main grids and work symbiotically to meet energy demands. Upon disruptions of power in main grids however, microgrids are able to disconnect and self-contain both energy supply and demand.

There are also rare examples of remote microgrids that do not have any connections to a central grid. The need for microgrids is often due to transmission line obstacles caused by financial limitations and/or geographic phenomena such as water bodies. El Hierro, one of the Canary Islands in Spain, as described by IRENA (2021) is a remote microgrid that has successfully integrated RESs that can meet 100% of the island's electricity demands. About 80% of local energy needs are met through a hybrid hydro–wind power system, and the remaining 20% is generated through solar power. In case of emergencies, El Hierro's former diesel generators remain in place. DE takes many forms at many scales in order to meet the many electricity needs of populations. The short- and long-term economic and environmental benefits of solar power, DH, and microgrids are proof that the investment into their respective infrastructures is necessary to ensure energy security in a carbon-neutral way moving forward.

Fig. 5.6 Microgrids often use various power sources such as solar power like the one shown here in Middlebury, Vermont, USA

5.3.1 Net-Zero Energy Systems

A popular term today within the discourse of energy generation and consumption is "net zero." According to the Government of Canada, a net-zero dwelling, or building, uses only as much energy as it can produce from on-site renewable energy (Canada, 2020) (Fig. 5.7). This principle can also be scaled up with goals of net-zero neighborhoods and cities. In fact, in June 2021, the *Canadian Net-Zero Emissions Accountability Act* came into effect enshrining the country's commitment to having net-zero GHG emissions by 2050 (Canada, 2022). All of the DE sources as mentioned above are steps in the right direction. Complementary to alternative energy sources are varying energy conservation methods, many of which are afforded through digital means. IoT-connected devices such as thermostats as mentioned in Chap. 1, programmable blinds, and app-connected lighting systems are all ways in which energy can be conserved to optimize use. Other methods include behavior changing and incentives such as household and office energy conservation games which use a mix of real and virtual game elements to educate and promote energy-conserving habits. These and other household motivators and automation devices will be further expanded on in Chap. 8.

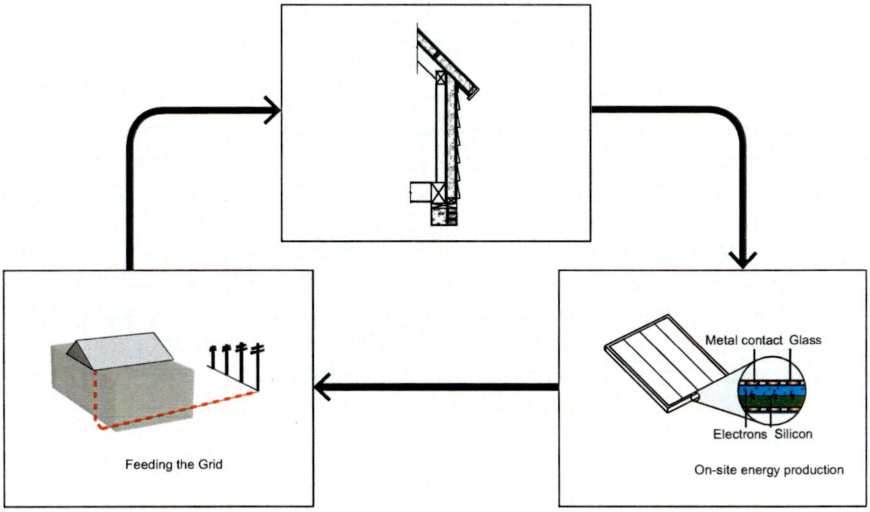

Fig. 5.7 Net-zero energy systems call for an increase in the energy efficiency of buildings, on-site energy production, and feeding the grid with excess power

5.4 Waste Management Using Digital Means

According to the World Bank (n.d.), 2.01 billion tons of municipal solid waste is produced annually, and at least 33% of it is not managed in environmentally responsible ways. By 2050, it is expected that annual waste generation will reach 3.40 billion tons. In terms of the composition of this waste, 44% is food, 17% is paper and cardboard, 12% is plastic, and other categories include glass, metal, rubber, leather, and wood. It is worth noting that these figures vary within and between nations based on income and other factors. The methods in which this waste is dealt with consist of 37% landfills, 33% open dump, and 11.1% incineration, among other techniques. This waste produces approximately 5% of annual global carbon dioxide (CO_2) emissions (World Bank, n.d.). While these immense amounts of global waste have the potential to cause problems, they also present opportunities for innovative solutions that promote a life cycle approach among the other principles of sustainability. Embedded within these solutions are employment opportunities, energy potential, and the reduction of GHGs. Waste management is a complex and expensive feat; therefore, this section is able to cover just the beginnings of the relevant digital solutions.

5.4.1 Eliminating Waste Before It Happens

The first step of tackling waste is to eliminate as much as possible. This can mean finding a second life for products by reselling or donating them; recall FoodRescue.ca in Canada covered in Chap. 2 that connects grocery stores and restaurants to nonprofit and social organizations. It can also mean sharing items with friends and family or through online libraries such as tool libraries. Many stores also rent out equipment so that it does not have to be purchased new. Another solution is to disallow products built for planned obsolescence, that is, products that are manufactured to wear out or become nonfunctional in a short period of time, forcing consumers to purchase a new product. As proposed by Satyro et al. (2018), this includes items designed purposely for fast consumption such as a cigarette lighter without flame control. A larger than necessary flame will be produced to light a cigarette making the consumer use the lighter faster than expected and therefore purchase a new one. The other subcategory of planned obsolescence according to Satyro et al. (2018) is electronics that are designed for restricted technological updates on either hardware or software. This looks like software developers opting not to support phones of a certain age, potentially blocking them from updates and forcing customers to purchase new phones or risk data and communication losses. Less waste means delaying obsolescence through technical and physical durability and replaceability.

In recent decades, planned obsolescence has been countered by the right to repair, a movement promoting the ability for bought items to be used, modified, and repaired as deemed fit by the owner. It also acknowledges the environmental detriment caused by the manufacturing and disposal of unnecessary items. Legislation surrounding the right to repair requires manufacturers to provide consumers of their products with information on how and where to repair them. This movement is supported by various grassroots organizations such as the Repair Café (n.d.) initiated in Amsterdam in 2007. Since its opening, the Repair Café has grown to more than 2200 locations worldwide. They provide free meeting places with tools, materials, and expert volunteers to which individuals can bring items in need of repair. Clothes, furniture, electrical appliances, and bicycles are among some of the many items that have successfully been repaired through Café services. Aside from the Repair Café, there are various online resources such as YouTube channels and company websites that provide information on how to and where to fix items in need of repair.

Another technique that supports waste diversion is taking a life cycle approach to ensure products and projects produce the least amount of waste possible (Fig. 5.8). This is supported by various digital tools such as the Ecolizer (n.d.), developed in Belgium to curb product waste when they are still in preliminary planning phases. It is a digital calculator geared toward designers and companies that estimates the environmental impact of a product based on its materials, processes, packaging, and transport modes. The environmental impact of each phase of the product's life cycle is determined for targeted changes as well as the product as a whole for comparison to others. Analyses have been performed on clothing, coffee makers, furniture, energy monitors, and more.

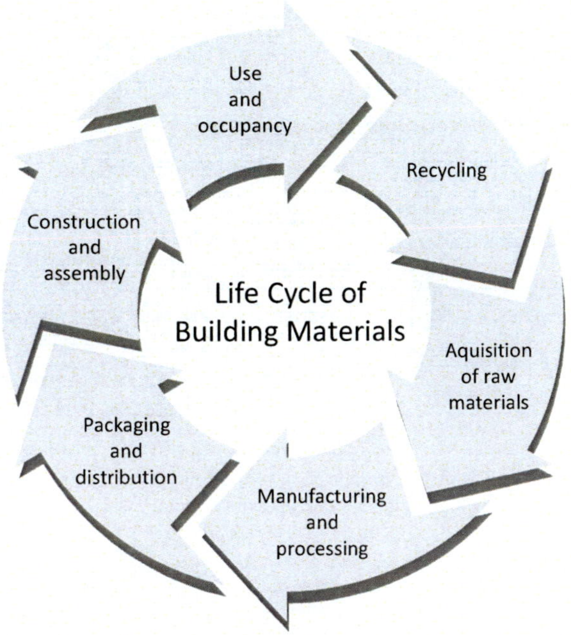

Fig. 5.8 A strategy that supports waste diversion is taking a life cycle approach to ensure that products and projects remain in use to produce the least amount of waste possible

5.4.2 Upcycling

Once it is established that waste from or of an item or project cannot be diverted, repaired, or given a second life without alteration, it should be determined whether or not the item can be upcycled. Upcycling is a form of recycling that involves the reuse of materials to create products of a higher value than the original material (Fig. 5.9). The materials used in these processes are often able to avoid entering municipal recycling systems, thus decreasing demand on these resources. The process of upcycling is largely aided by online resources aimed at connecting discarded materials with hopeful users. Websites such as Freecycle.org, Reuse Marketplace, and Reusewood.org all provide services connecting those with unwanted building materials and more to potential buyers, traders, or donation receivers. Other online tools such as Pinterest provide ideas of what and how to upcycle household objects that would otherwise be wasted.

The Sea Train Residence in Los Angeles, California, is an example of upcycled commercial and industrial waste used to create a unique urban home (Green Design, 2022). The dwelling uses shipping containers for the master bedroom and grain trailers for an exterior lap pool and koi pond. Other reclaimed materials include steel beams and cladding, wood joists, and glass, some of which were salvaged on-site. The reuse of local material means little or no transport and lower waste, time, and labor costs.

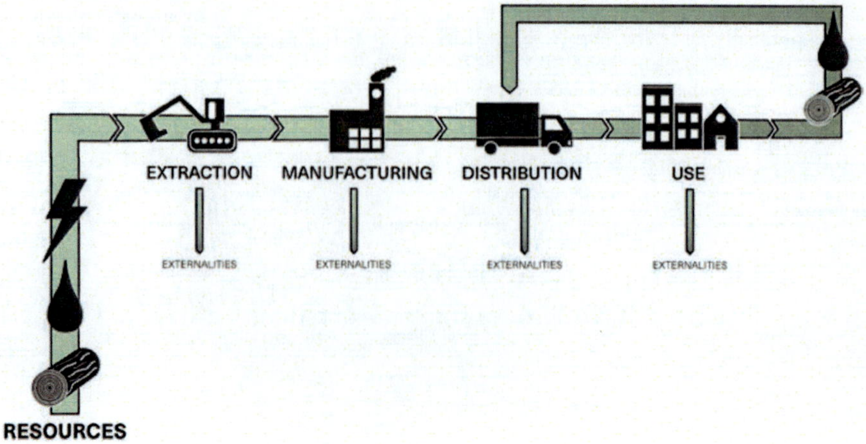

Fig. 5.9 Upcycling is a form of recycling that involves the reuse of materials to create new products of a higher value than the original material

Fig. 5.10 Shipping containers are used as prefabricated student housing in Amsterdam, the Netherlands

The Keetwonen project in Amsterdam, similarly, featured upcycled material by using over 1000 used shipping containers to develop 1034 prefabricated units of student housing (Martinez, 2017). This temporary project was an attempt to tackle the lack of student housing in Amsterdam (Fig. 5.10). Each unit contains a large window, private bathroom, and a bed. Laundry rooms, coffee shops, and other communal areas are also featured within these buildings. Originally slated as a temporary installment, these units were so successful that they have since remained, with some being moved to create more student housing at Groningen University, also in the Netherlands. While Keetwonen on the one hand offers dense affordable housing to students, it also took a production line in China to make them. This allowed for rapid production of units, around 40 per week, though the environmental and economic transport costs seem to be in direct conflict with the benefits.

5.4.3 Making Waste Management and Collection Efficient Using Digital Means

Despite all the above techniques, the world is still plagued with billions of tons of solid waste. Because of this, postconsumer waste management sees products being landfilled, incinerated, recycled, composted, or anaerobically digested, among other techniques. While these systems have their flaws, many are also presently accompanied by various technological solutions. For example, the decomposition of materials in landfills creates GHGs and leaches toxic pollutants into nearby soils and water bodies. Protective liners are one solution to limit groundwater contamination of leachate but are a proactive solution that is not able to address the present landfill situation (Fig. 5.11). Acting retroactively are various monitoring techniques that can identify sites or objects in landfills emitting excessive methane—which typically makes up 50% of landfill gas (Henkelman & Sullivan, 2019). Surface emission monitoring (SEM) techniques, for example, use portable methane meters to measure concentration throughout landfills. On a larger scale, infrared (IR) cameras allow for methane to be detected both from the landfill itself and from leaks in collection and control systems. IR cameras can be handheld, mounted, or drone-mounted extending monitoring to inaccessible areas. The monitoring of emissions

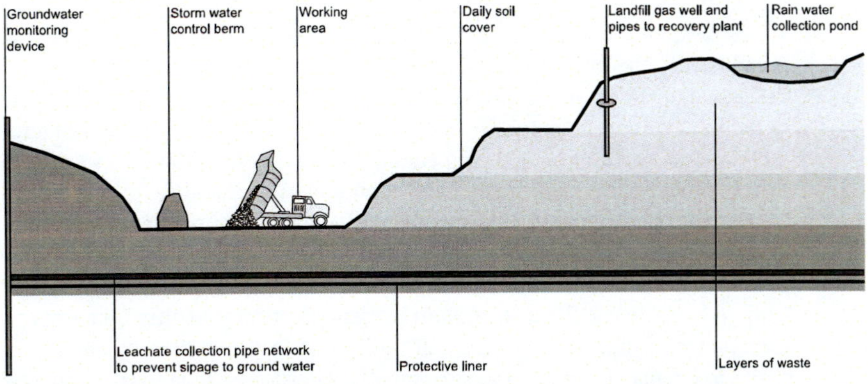

Fig. 5.11 Protective liners are used to limit groundwater contamination in landfills

allows for action to be taken when methane levels risk causing explosions and/or health hazards or when unnecessary methane leaks appear.

Recycling also poses a variety of issues in terms of feasibility. Each type of plastic must be processed separately; even attached lids and bottles made from different plastics must be taken apart. This means people and machines must separate any recycling that makes it to a treatment facility. After sorting, items are washed, shredded, melted, and remolded. If the wrong plastics are melted together, however, the final material will be worthless. One solution to this is by introducing "compatibilizers," chemicals that help different plastics blend. While there isn't a one-size-fits-all compatibilizer, Cornell chemist Geoffrey Coates and his team have developed one that combines polyethylene and polypropylene—two plastics that make up a large portion of the world's waste (Temming, 2021). The development and introduction of effective compatibilizers will cut back the amount of time, space, and energy that goes into sorting plastics as well as allow for more types of plastic to be recycled.

Waste collection methods can also be addressed with technological solutions. The processes of garbage, recycling, and compost trucks making daily city rounds to collect waste have transport trucks polluting excess GHGs and can be made far more efficient. In Hammarby Sjöstad, a "neighborhood living lab" in Stockholm, Sweden, for example, the need for waste pickup services has been greatly reduced (Fig. 5.12). A pressurized air transport system moves waste through pipes to central collection points, drastically reducing the number of stops made by pickup trucks and reducing GHGs produced in the process (Friedman, 2017) (Fig. 5.13). Another improvement in terms of waste collection is the application of IoT devices (Ali et al., 2020). Sensors can be placed in both residential and public waste bins to track how full they are. This information is transmitted to a central controller that can create routes for waste collection vehicles that prioritizes the fullest bins. GIS technologies, such as those discussed in Chap. 9, integrated into these networks provide the added ability of optimizing routes to ensure vehicles are collecting the most amount of waste while emitting the least amount of GHGs.

Lastly, GIS also allows for the least detrimental locations for all types of waste management plants. Being able to account for topography, dwellings, wind patterns, and emissions, among other factors, in a single analysis means that impact on humans and the environment of waste treatment can be kept to a minimum. As described, there are many different applications of new and developing technologies to the sector of waste management. While an emphasis on waste elimination is key, technologies that optimize the processing of inevitable leftover waste play key roles in achieving both environmental and economic sustainability.

As presented, there are various technological solutions that better allow for the energy and waste management needs of today's growing population to be met in a sustainable manner. Technologies that integrate RESs better manage the electricity grid and promote DE act to support present electricity habits. Likewise, technologies that optimize waste collection and mitigate the effects of already-produced waste act to manage waste more efficiently. These approaches are complemented by various proactive waste-elimination techniques presented in this chapter: sharing,

Fig. 5.12 Hammarby Sjöstad is a mixed-use neighborhood covering 200 hectares (494 acres) and includes apartments, stores, offices, and extensive green spaces

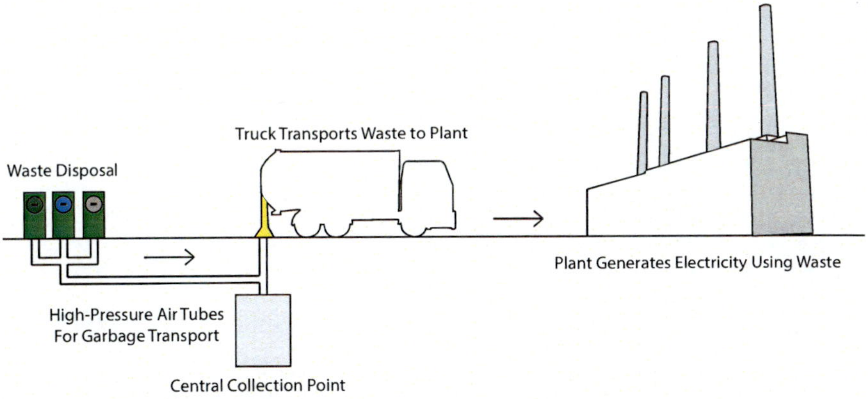

Fig. 5.13 A pressurized air transport system in Hammarby Sjöstad moves waste through pipes to central collection points, drastically reducing the number of stops made by pickup trucks and reducing GHGs produced in the process

renting, reselling, repurposing, repairing, and better building, all of which are supported by digital means. The role technology plays in supporting the different types of sustainability is unfounded; it cannot go unsaid however similar to other chapters that consumption habits must also change in order to ensure resources for future generations. Later, Chap. 8 will introduce the role technology can play in changing consumption habits to round out the solution to the energy crisis.

Questions for a Follow-Up Discussion

1. What is a smart grid, and can its efficiency be improved using digital means?
2. What is DE, and can it contribute to achieving energy efficiency?
3. What are the attributes of DH that lead to having an energy-efficient system?
4. What are net-zero neighborhoods and homes, and how are they designed?
5. How can efficient waste management systems be achieved using digital means?

References

Abnett, K., & Jessop, S. (2021). "U.S., Canada among 20 countries to commit to stop financing fossil fuels abroad". Reuters. https://www.reuters.com/business/cop/19-countries-plan-cop26-deal-end-financing-fossil-fuels-abroad-sources-2021-11-03/. Accessed April 14 2022.

Ali, T., Irfan, M., Alwadie, A. S., & Glowacz, A. (2020). IoT-based smart waster bin monitoring and municipal solid waste management system for smart cities. *Arabian Journal for Science and Engineering, 45*. https://doi.org/10.1007/s13369-020-04637-w. Accessed April 14 2022.

Bollen, M. H. (2017). "The power grid as part of a 100% renewable energy system". Large Scale Grid Integration of Renewable Energy Sources. Chap 1, pp 1–27. IET Digital Library. https://doi-org.proxy3.library.mcgill.ca/10.1049/PBPO098E_ch1. Accessed April 14 2022.

Canada. (2017). "About Renewable Energy". https://www.nrcan.gc.ca/our-natural-resources/energy-sources-distribution/renewable-energy/about-renewable-energy/7295. Accessed April 14 2022.

Canada. (2020). "Net-Zero: future building standards". https://www.nrcan.gc.ca/energy-efficiency/energy-efficiency-homes/buying-energy-efficient-new-home/netzero-future-building-standards/20581. Accessed April 14 2022.

Canada. (2022). "Net-Zero Emissions by 2050". https://www.canada.ca/en/services/environment/weather/climatechange/climate-plan/net-zero-emissions-2050.html. Accessed April 14 2022.

Center for Climate and Energy Solutions. (n.d.). "Global Emissions". https://www.c2es.org/content/international-emissions/#:~:text=Most%20of%20the%20world's%20greenhouse,the%20United%20States%20and%20Russia. Accessed April 14 2022.

Clean Energy Grid. (n.d.). "Our Outdated Grid". Americans for a Clean Energy Grid. https://cleanenergygrid.org/our-outdated-grid/. Accessed April 14 2022.

Creative Energy. (n.d.). "District Energy". http://creative.energy/district-energy. Accessed April 14 2022.

DEC. (n.d.). "What is DE?". Decentralized Energy Canada DEA. https://www.deassociation.ca/whatisde. Accessed April 14 2022.

Ecolizer. (n.d.). https://www.ecolizer.be/. Accessed April 14 2022.

Enbridge Gas. (n.d.). "Renewable Hydrogen". https://www.enbridgegas.com/sustainability/renewable-hydrogen. Accessed April 14 2022.

Energy.gov. (n.d.). "Solar Integration: Inverters and Grid Services Basics". Office of Energy Efficiency & Renewable Energy. https://www.energy.gov/eere/solar/solar-integration-inverters-and-grid-services-basics#:~:text=An%20inverter%20is%20one%20of,which%20the%20electrical%20grid%20uses. Accessed April 14 2022.

EPA. (2017). "Geothermal Energy". Environmental Protection Agency EPA. Archive. https://archive.epa.gov/climatechange/kids/solutions/technologies/geothermal.html

Friedman, A. (2017). *Designing sustainable communities*. Bloomsbury.

Friedman, A. (2022). *Designing innovative sustainable Neighbourhoods*. Routledge.
Green Design. (2022). "Sea Train House". Northern Architecture. https://www.northernarchitecture.us/green-design-2/sea-train-house.html. Accessed April 14 2022.
HELEN. (n.d.). "Helsingin Energia's smart CHP/DH system—the most energy-efficient solution for heating Finland's capital". Helsingin Energia. https://www.districtenergyaward.org/wp-content/uploads/2012/10/Modernization_Finland_Helsinki_2011.pdf. Accessed April 14 2022.
Henkelman, J., & Sullivan, P. (2019). "Understanding landfill gas monitoring techniques". Waste Today Magazine. https://www.wastetodaymagazine.com/article/landfill-gas-monitoring-techniques/. Accessed April 14 2022.
IEA. (n.d.). "Renewables". International Energy Agency IEA. https://www.iea.org/fuels-and-technologies/renewables. Accessed April 14 2022.
IESO. (n.d.). "A Smarter Grid". Independent Electricity System Operator. https://www.ieso.ca/en/Learn/Ontario-Power-System/A-Smarter-Grid. Accessed April 14 2022.
IRENA. (2015). "Renewable Energy Integration in Power Grids". International Renewable Energy Agency. Abu Dhabi. https://www.irena.org/-/media/Files/IRENA/Agency/Publication/2015/IRENA-ETSAP_Tech_Brief_Power_Grid_Integration_2015.pdf. Accessed April 14 2022.
IRENA. (2021). "Renewable Power Generations Costs in 2020". International Renewable Energy Agency, Abu Dhabi. https://www.irena.org/-/media/Files/IRENA/Agency/Publication/2021/Jun/IRENA_Power_Generation_Costs_2020.pdf?la=en&hash=755CB6E57667D53B629967EC7F9BE57A55D1CD78. Accessed April 14 2022.
Kahan, A. (2019). "EIA projects nearly 50% increase in world energy usage by 2050, led by growth in Asia". U.S. Energy Information Administration EIA. https://www.eia.gov/todayinenergy/detail.php?id=41433
Lichtenwoehrer, P., Erker, S., Zach, F., & Stoeglehner, G. (2019). Future compatibility of district heating in urban areas—A case study analysis in the context of integrated spatial and energy planning. *Energy, Sustainability and Society, 9*(1), 1–12.
Martinez, O. (2017). "Keetwonen, Amsterdam's Cargo Container Student Housing Complex". Containerpedia. https://www.containerpedia.com/keetwonen-amsterdam-s-1000-cargo-container-student-housing-complex. Accessed April 14 2022.
NEA. (n.d.). "Geothermal". National Energy Authority of Iceland". https://nea.is/geothermal/. Accessed April 14 2022.
Project Solar UK. (2018). "The History of Renewably Energy: Where It All Began". https://www.projectsolaruk.com/blog/history-renewable-energy-began/#:~:text=Renewable%20energy%20%E2%80%93%20defined%20as%20energy,mimic%20the%20workings%20behind%20hydropower. Accessed April 14 2022.
Repair Café. (n.d.). https://www.repaircafe.org/en/about/. Accessed April 14 2022.
Satyro, W., Sacomano, J., Contador, J., & Telles, R. (2018). Planned obsolescence or planned resource depletion? A sustainable approach. *Journal of Cleaner Production, 195*, 744–752. https://doi.org/10.1016/j.jclepro.2018.05.222. Accessed April 14 2022.
Shively, B. (n.d.). "How Much Primary Energy is Wasted Before Consumers See Value from Electricity?". Enerdynamics. https://www.enerdynamics.com/Energy-Currents_Blog/How-Much-Primary-Energy-Is-Wasted-Before-Consumers-See-Value-from-Electricity.aspx. Accessed April 14 2022.
SmartGrid. (n.d.). "The Smart Grid". U.S. Department of Energy. https://www.smartgrid.gov/the_smart_grid/smart_grid.html. Accessed April 14 2022.
Student Energy. (n.d.). "Electrical Grid". https://studentenergy.org/distribution/electrical-grid/
Temming, M. (2021). "New recycling technologies could keep more plastic out of landfills". Science News for Students. https://www.sciencenewsforstudents.org/article/new-recycling-technologies-could-keep-more-plastic-out-of-landfills. Accessed April 14 2022.
UNFCCC. (n.d.). "The Paris Agreement". United Nations Climate Change UNFCCC. https://unfccc.int/process-and-meetings/the-paris-agreement/the-paris-agreement. Accessed April 14 2022.

United States. (2021a). "Electricity explained". U.S. Energy Information Administration EIA. https://www.eia.gov/energyexplained/electricity/delivery-to-consumers.php. Accessed April 14 2022.

United States. (2021b). "Renewable energy explained". U.S. Energy Information Administration EIA. https://www.eia.gov/energyexplained/renewable-sources/

United States. (n.d.). "Benefits of Residential Solar Energy". U.S. Department of Energy. https://www.energy.gov/energysaver/benefits-residential-solar-electricity. Accessed April 14 2022.

Vivint Solar. (n.d.). "What is net metering?". https://www.vivintsolar.com/learning-center/what-is-net-metering. Accessed April 14 2022.

Wood, E. (2020). "What is a microgrid?". Microgrid Knowledge. https://microgridknowledge.com/microgrid-defined/#:~:text=A%20microgrid%20is%20a%20self,generators. Accessed April 14 2022.

World Bank. (n.d.). "What a Waste 2.0". The World Bank. https://datatopics.worldbank.org/what-a-waste/trends_in_solid_waste_management.html#:~:text=The%20world%20generates%202.01%20billion,from%200.11%20to%204.54%20kilograms. Accessed April 14 2022.

Chapter 6
Knowledge Transfer, Social Engagement, and Culture

New technological solutions provide cities with the ability to safeguard their children's educations and animate their streets to promote cultural sustainability. Better data collection methods and analysis software means that city planners and municipal governments can study the patterns in their education system to pinpoint areas for improvement. With the recent popularization of remote learning, just because a child cannot physically be present in school does not mean that they need to miss that day's teachings. There are also opportunities for formative learning experiences for everybody outside the classroom through public art or cultural events. The ways in which we learn and pass on knowledge have been nothing short of transformed by communication technologies in the twenty-first century. With these changes come new opportunities to design communities for sustainability. This chapter discusses and profiles the ways that technology has revolutionized three different dimensions of intellectual life—social media, education systems, and cultural institutions—and analyzes case studies that demonstrate how we can leverage technology to design a more sustainable world.

6.1 Past and Present Knowledge Transfer

The way contemporary knowledge is transferred is much different from how it has been historically. Throughout history, knowledge has been rather location based, with little means of communication. Communities around the world gathered their own novel and necessary information for survival, entertainment, and celebrating their culture (Fig. 6.1). Open space devoted to public gathering and information exchange has formed an integral part of the urban and cultural heritage of many societies throughout history and played a critical role in the genesis of commerce, the emergence of democracy, and the vitality of civic life. Bearing different names, the square is known as the *Agora* in ancient Greek towns, the *Plaza* in cities of

A. Friedman, *The Sustainable Digital City*,
https://doi.org/10.1007/978-3-031-25488-8_6

Fig. 6.1 Cultural traditions are celebrated in Barcelona, Spain, and in Valletta, Malta, neighborhoods

Fig. 6.2 Open space devoted to public gathering and information exchange like this Greek Agora has formed an integral part of the urban and cultural heritage of many societies throughout history

Spanish origins, *Piazza* in Italy, *Village Green* in settlements with feudal pasts, and *Market Square* in others (Fig. 6.2). People's comings and goings were reported by word of mouth or, later, the telephone when invented. Music and art were experienced firsthand at shows, get-togethers, and cultural ceremonies. Hands-on and academic skills were taught through mentorship, stories, and books. Presently, we have access to all of these knowledge-transferring circumstances as well as a plethora of new formal and informal ones, many afforded through technology. Information is no longer unique to its sourced location but easily accessed and shared globally through the Internet and connecting devices. Social media websites such as Facebook, Twitter, and Instagram quickly gained worldwide attention and are now dominant sources of news and other information, not to mention the newspapers and magazines, more formal information sources that are mostly available through digital means as well.

It seems that as we progress through the twenty-first century, the reach of technology is furthering itself into our knowledge cultures. One needs to look no further than the fact that *Googling* (now a common verb) is the touchstone of common information discovery for any knowledge seeker to see how the paradigm has changed. With mobile technologies achieving incredible penetration into markets across the world, search-result algorithms and social media feed algorithms now determine to a great extent what we know and when we know it. This is to say nothing of sensors and monitors that now enliven our cities and create an Internet of things (IoT). These Internet-enabled devices can provide real-time feedback and are being integrated into our education systems and cultural institutions for all manner of purposes, from monitoring air quality to protecting the health of children to sensing our concentration at work. Even changes such as the switch from analog to digital billboards allow for much more information such as job opportunities, tourist attractions, sales at local stores, historical information, and more to be conveyed to passersby. If you can, imagine the billboard scene in New York City's Times Square and how impactful those huge neon signs are at delivering information (Fig. 6.3). These changes are occurring so rapidly that it hardly feels as though we can direct these waves of innovation. Yet, design interventions among other approaches can help us leverage the exciting affordances of these new technologies to create a more sustainable future.

6.2 The Rise of Social Media

Social media has become a central component of twenty-first century's intellectual life. Anyone hoping to keep up with current events, track political messages, connect with cultural institutions, or even find political and cultural events in which to participate finds themselves drawn into the social media universe by network effects: since so many other people use social media, to stay "in the know," an individual must also use social media. The short video-sharing app TikTok has become increasingly popular in recent years, reaching over a billion users and acting as a hub for youth culture (Smith, 2021). Part of its popularity comes from its video-generating algorithms which are known to be spot-on at reading and predicting user preferences. It relies on likes, comments, retention (whether a user comes back), and time spent (how much time is spent on any type of video to cater to user taste) (Smith, 2021). Because of this, users are fed specific variants of video types they engage with most on the app from cute animal video clips down to political rabbit holes.

Fig. 6.3 The switch from analog like the ones shown here in New York City to digital billboards allows for much more information to be displayed such as job opportunities, tourist attractions, and sales at local stores

One of the urban systems social media has impacted is the way that individuals participate in civic life in many ways, for one, in its governance. In the early days of social media, much was made of the average citizen's ability to communicate directly with local government. McCLellan et al. (2018) argue that social media becomes the most used channel for citizens to communicate to the government. Townsend (2014) suggests that for a new cadre of civic leaders, smart technology is not just a way to do more with less. It is a historic opportunity to rethink and reinvent government on a more open, transparent, democratic, and responsive model.

Indeed, since the dawn of social media, conventions for the management of these "direct" channels have been established, and they are now quite integrated into local life. Viral Twitter and Facebook posts are regularly the subjects of broadcast news stories and often mark the beginning or even make up the substance of entire political debates or tensions, be they interactions between local councilors, photos of public infrastructure that needs fixing, or videos of events that spark social movements.

Relatedly, social media has altered governance in that it has changed the way in which citizens connect with one another to coordinate spontaneous, openly accessible events, the key being that social media can allow individuals who do not know one another closely to very quickly disseminate information. These events include so-called *pop-up* cultural experiences, from restaurants to musical performances and, of course, political rallies and demonstrations (Fig. 6.4).

How can these new ways of circulating information be leveraged to create sustainable communities? One method is by integrating social media into environmental assessment processes. According to Sinclair et al. (2017), impact assessment is the process of considering the economic, social, and environmental ramifications of a proposed project. Public participation is the center of this process (Devlin et al., 2005). Public participation exemplifies sustainable goals and values, given that it promotes deliberative democracy, collaborative rationality, and environmental justice (Morgan, 2012). Given contemporary communication practices, what better way to orchestrate public participation in environmental impact assessments (IAs) than through social media? As previously mentioned, social media has become a conventional means by which citizens communicate with local government but also how they communicate with one another to organize in response to proposed projects. Social media can be used in environmental IAs to promote both forms of communication.

Sinclair et al. (2017) suggest that in the IA context, social media has been useful for uniting the voices of individuals who have the same views on planning and development issues and helping them to organize to participate in IA cases. Social media can be instrumental in gathering signatures for petitions that can be a part of

Fig. 6.4 Social media allowed individuals who do not know one another closely to very quickly disseminate information about events to including so-called pop-up cultural experiences

environmental assessment processes. For instance, the World Wildlife Fund used a social media campaign to gather petition signatures for a Liquefied Natural Gas terminal site project in Hong Kong. As a result, over 20,000 petition signatures were submitted as part of the formal consultation process (Sinclair et al., 2017).

In short, social media should be regarded as a rapid response "opinion-amplifier" that is able to quickly bring together public voices in a more effective way than conventional media, such as print, TV, and radio news (Sinclair et al., 2017). Yet, there are drawbacks in the utilization of social media for public consultation processes. As stated at the International Association for Impact Assessment's 2016 conference, it is important that participants may feel that, through their cursory online involvement, they are "voting" for a preferred outcome instead of truly participating in the process (IAIA Session Discussion, 2016). This may be unfortunate because many formal processes for participation still require channels outside of social media. The central legal mechanism is still the public hearing, which requires a separate registration process. If citizens do not participate in this way, their voices may not be heard (Evans-Cowley & Hollander, 2010).

That said, another exciting approach to leveraging social media may be to utilize social media in conjunction with gaming technologies to reach younger demographics as part of consultation processes. Local government in Hong Kong has utilized the popular city-building computer games SimCity and Megapolis toward these ends (EPD, 2015). The models on display are interactive, so that the public can make use of the tools to offer suggestions, alternative options, and mitigation measures (EPD, 2015). This allows for a new depth of citizen participation in consultation processes. More on digital citizen participation and inclusive governance can be found throughout Chap. 2.

Another one of the many impacts social media has had in our lives is the ability to adapt culturally when moving to a new place and address cultural differences. Chen and Sawyer (2012) in their article *The Impact of Social Media on Intercultural Adaptation* explore how people use social media to better integrate themselves into host cultures following migration. As globalization progresses, people migrate around the world for education, work, family, and humanitarian reasons. The intercultural adaptation processes that follow migration are honeymoon, crisis, adjustment, and biculturalism; one might imagine what these stages consist of (2012). By speaking with international students at American universities, Chen and Sawyer found that migrants can use social media to strengthen, build, and maintain relationships (2012). Interviewed students were also able to familiarize themselves with culture beforehand through social media and maintain relationships with friends and family back home.

Presently, videos circulating TikTok, Facebook, YouTube, and more depict culture shocks people have after moving to a new country. A culture shock is according to Kohls (2001) the psychological disorientation most people experience when they move for an extended period of time into a culture markedly different from their own. Take American travel consultant Kacie Rose who recently moved to Italy (Rose, n.d.). She has been documenting her personal experience with cultural differences between the United States and Italy such as her unfamiliarity with the Italian schooling systems, drinking cultures, and healthcare (n.d.). As culture and migration become increasingly global, social media provides a space for people to learn about and discuss differences, similarities, and experiences among cultures. There is a responsibility however to social media users and the sites themselves to monitor discussions and ensure they are respectful and appreciative. There have been instances, such as the 2021 Capitol Hill riot in Washington DC, where social media has been used as a coordination tool for violent events.

To sum up in the dawn of social media, the ability of these tools to circulate information in new ways was greeted with unbridled optimism, given that the technological affordances of these new methods created the potential for systems of unprecedented transparency and breadth. Yet, this techno-utopianism has soured as the unintended consequences of such tools have become apparent. A lack of oversight and transparency of tech giants has meant that these extremely powerful tools, upon which public intellectual life relies, are not in the hands of the public. Whether these tools remain useful to those who disagree with corporate or government interests is greatly limited by the matter of whether or not there emerge accountability mechanisms for these giants over the course of the next decade.

6.3 Technological Change and Education

Technology has been transforming education, especially in the area of remote learning which became apparent during the COVID-19 lockdowns. Sustainability also means resiliency in emergency, and therefore pandemic preparedness has emerged as a part of sustainable thinking. Students packed into larger and larger class sizes are simply unsuitable in a time when transmission of a deadly virus is easily possible. MOOCs (massive online open courses) such as those offered by Open University, as discussed in Chap. 4, have been explored to support postsecondary education for decades. Other institutions like the nonprofit Khan Academy have aimed their sights at aiding those at the elementary and high school levels for over a decade; however, it was not until many learners were forced remote following the onset of the COVID-19 pandemic that there have been large-scale social adaptations to the issue of remote learning. Two main tactics to explore how to sustainably adapt to this new world are attempting to close the digital divide by providing Internet to those without access and retrofitting or building new structures to support pandemic-safe learning.

6.3.1 Internet Infrastructure to Support Remote Learning

One major way in which cities have aimed to become more sustainable and support remote learning is simply through attempting to close the digital divide. The digital divide, which is touched on throughout this novel, runs deep, especially along rural–urban and income lines. Those in rural areas and with less income are far less likely to have high-quality access to online content (Turner-Lee, 2020a). In the United States alone, 21.3 million people lack broadband access (Turner-Lee, 2020a). Usually, many children will use libraries or commercial establishments that offer free Wi-Fi in order to close the gaps in their resources at home, but this is simply not possible when these institutions are closed due to public health orders to promote social distancing (Turner-Lee, 2020b). In particular, broadband services are vital because they allow students to consume all-important video content. Video content is the heart and soul of remote connections between students and teachers and vitally also to peers, as it allows for a synchronous social connection that supports motivation and community-building (Turner-Lee, 2020b). This is to say nothing of the fact that a lot of learning material can be delivered in video format by both public schools and institutions like Khan Academy.

Some school boards have adapted to our present era with great agility in a brilliant way: utilizing school buses. During the COVID pandemic, school buses are no longer required to frequently physically transport students to classroom learning environments and so lay empty and unused. A solution actually pioneered pre-pandemic by the rural Coachella Valley Unified School District in California has become more popular: outfit school buses with solar-powered Wi-Fi routers, and park them in underserved areas so that kids can have more equitable access (Turner-Lee, 2020a). This brilliant adaptation has been called *Wi-Fi on Wheels* (Turner-Lee, 2020a) and has been used in Topeka Public Schools in Topeka, Kansas, Hoover, Alabama, Wisconsin, the Mosinee School District, and also South Carolina (Al-Arshani, 2020).

6.3.2 Building Infrastructure to Support Pandemic-Safe Learning

Architecture firms are quickly adapting to the new normal and creating best new practices for buildings in a speedy response to our changing world. As explored by the firm Gow Hastings Architects (2020) in Canadian Architect Magazine, buildings can be retrofitted to adapt to remote learning. One vital element of this process is integrating audiovisual tech into classrooms, that is, the construction of places where lectures can easily and conveniently be recorded and streamed to students. This means that these rooms must contain equipment, convenient and versatile electrical power options, and even green screens and soundproofing. Further, designers can configure classrooms to provide adequate space, simply by integrating flexibility as a means to transform spaces (Fig. 6.5). Recessed floor outlets do not constrain furniture outlets in the same ways as those that emerge from floors or walls. An emphasis can be put on utilizing spaces in a building that are square as classroom spaces. That is because these spaces can more easily be accommodating for different and distanced layouts, as opposed to more inflexible long and narrow spaces.

Fig. 6.5 Designers can configure classrooms to provide adequate places for unique functions by using flexible partitions like the one shown here to transform a space

The advantage of this approach is that as pandemics ebb and flow in severity, these spaces can adapt, too, and be reconfigured. Though prompted by the COVID-19 pandemic, these adaptations, aside from allowing for safe and efficient remote or physically distanced learning, generally increase the accessibility of learning opportunities when there is not a pandemic at play.

Gow Hastings (2020) has social sustainability in its recommendations for designing interior spaces suitable for distance learning. Their intention is to ensure that social distancing in classrooms does not stifle collaboration and connection. We design learning spaces that promote the cross-pollination of ideas. This means finding clever ways to nudge or remind students about social gathering through design but striking a balance wherein they can still communicate and collaborate. These architects offer using fixed tables and intentionally designing seating arrangements to discourage large groups.

6.4 Cultural Institutions

While the purpose of many public cultural institutions is to keep society connected to traditional forms of knowledge, they have not been untouched by technological change. In fact, these institutions have been innovating rapidly to showcase art and knowledge by way of technology. Take, for example, the many public art and cultural events that utilize technology (Fig. 6.6). Art installations like Montreal's Cité Memoire project videos on the side of historic buildings to offer onlookers a glimpse into the city as it was decades prior. While this installation livens up plain walls, it also educates bystanders of the city's history which helps to foster a sense of pride and community. It is a unique unification of past and present. One can also take into account the various exhibitions that now include augmented reality components. When visiting a historic site, access to Wi-Fi and knowledge about a place are also available using a downloadable app (Fig. 6.7). The Art Gallery of Ontario's recent exhibition The Anthropocene included such elements to incredible effect. When visitors downloaded the Art Gallery of Ontario's mobile application, they could view endangered species as interactive digital presence in gallery space, for instance, an extinct rhinoceros. Communicating environmental loss through this medium added to the haunting atmosphere for the show's underlying message. Another example is the variety of webinars offered often by public or private entities such as city libraries, community centers, or universities and facilitated by virtual meeting services such as Zoom or Teams. The Toronto Public Library has numerous different online programs offered such as coding workshops, "how to's," author talks, and book club discussion groups (TPL, n.d.). While many of these sessions offered are hardly new in their conception, the digital element allows for increase accessibility

Fig. 6.6 Public art and cultural events like the Frida Kahlo immersive exhibit utilize technology to enhance civic life engagement

Fig. 6.7 When visiting a historic site like this one in Alberobello, Italy, access to Wi-Fi and knowledge about a place is available using a downloadable app

on the side of both hosts and participants. There is also the novel ability to host speakers and other guests from around the world.

Many museums, libraries, and universities, among other cultural institutions, exist under the premise of having unique information. This is no longer the case however, as online and often free sources of information grow in size and reach. Take Encyclopedia Britannica, the longest running in-print encyclopedia in the English language (Wikipedia, 2022a). It was first printed in 1771 but since 2011 has and will be published online exclusively. Relatedly, Wikipedia, the self-proclaimed "free encyclopedia that anyone can edit," has only ever been published electronically (Wikipedia, 2022b). The Wikipedia site has 2 billion unique device visits per month and is home to more than 58 million articles ranging in topics from culture and arts which constitute 30% of content to mathematics and logic 1% and everything in between: biographies and persons 15%, history and events 11%, technology and applied science 4%, and more (2022b). In 2014, the website was ranked the fifth most visited website following Yahoo, Facebook, Microsoft, and Google. It has since fallen to the 13th rank trailing behind many growing online shopping websites. Wikipedia has articles published in 326 languages, a testament to its global reach and the validity of crowdsourced information. For accuracy and safety however, the site's volunteer editor community is tasked with the responsibility of fact-checking contributed information and protecting (restricted editing) on controversial, sensitive, or vandalism-prone pages as deemed necessary (2022b). These new facets of knowledge transfer do not make cultural institutions redundant but force them to find new ways to facilitate both in-person and online learning, as will be explored in this section.

6.4.1 The Living Room Libraries of the Future

With the advent of new communication technologies, libraries have been under significant pressure to innovate. John Palfrey, author of *BiblioTECH* (2015)*: Why Libraries Matter More Than Ever in the Age of Google*, suggests that the original design of libraries—as glorified storehouses for books and manuscripts and pleasant places to read them—no longer suffices. People have many more options today to get access to knowledge (Fig. 6.8). The job of librarians is rapidly changing in this new world. Librarians have been rising to this challenge, and new best practices have emerged. The goal now, in the words of information leader Jeffrey Schnapp of Harvard's metaLAB, is for librarians to create a hybrid space where analog and digital worlds can coexist.

The struggle of Librarians to create programs and resources for these new types of spaces has of course been echoed in the struggle of architects to design for these new activities. Again, Palfrey offers significant insight and suggests that librarians—and the architects of libraries, for that matter—are grappling seriously with

Fig. 6.8 The original design and use of libraries—as glorified storehouses for books and manuscripts and pleasant places to read them like this one in Stuttgart, Germany—no longer suffices. Today, people have many more options to get access to knowledge including digital ones

the connection between physical architecture and information architecture. This means that more and more often, libraries are serving two purposes: traditionally contemplative spaces, quiet and appropriate for study. Palfrey also suggests that the contemplative spaces in libraries are well worth preserving, in part because they are lovely and in part because the always-on, highly connected pace of digital-era life can be overwhelming. As a public "living room," they can be warm, homey, and a place for social interaction, group work, and company. According to the author, the spaces of learning can demand the support and camaraderie that other humans provide.

As described by *The Globe and Mail* during the period of construction, the city of Halifax, Canada, is presently transforming its Central Public Library from a dusty book warehouse to a community hub for the digital age (Georgijevic, 2016). Bruce Gorman, then the director of the Central Library and Regional Services, aimed to disrupt the idea that libraries could not be social hubs. He aimed to offer not what people expect to see when they are going into a public library. We are shifting people's thoughts about what public libraries are (Taber, 2014). Indeed, the architects conceptualized it as a "city's living room" (Georgijevic, 2016). That is, while it offers some traditional, contemplative quiet spaces, it provides hubs for activities that one would never associate with a library before the Internet age: a 300-seat performance space, gaming stations, two cafes, and music stations at which a performer could burn a CD on the spot, a boardroom for use by local small businesses, seminar rooms, dedicated spaces for First Nations reading circles, and more (Georgijevic, 2016).

This new highly social purpose for libraries is why these buildings have become a major part of what can make a forward-thinking city to be simultaneously technologically relevant and sustainable. These institutions become a meeting place, allowing intergenerational interaction between all walks of life. Indeed, this is reflective in many elements of the Halifax public library's design, what with "the building's dramatic cantilevered glass-encased fifth storey features an open space for socializing, with descending stairs doubling as seating, reminiscent of a contemporary agora" (Georgijevic, 2016). It was designed by Morten Schmidt's firm Schmidt Hammer Lassen and also Nova Scotia-based partners, Fowler Bauld & Mitchell, with this exact purpose in mind. Schmidt suggests that a public library is probably as important as a church today or even more important than a city hall because it is the people's building and everyone can come there (Georgijevic, 2016).

The Calgary Central Library is a similar example (Fig. 6.9). In the words of Architect Vanessa Kassabian, "Libraries now are almost more social spaces than they are about learning" (Georgijevic, 2016). This philosophy is certainly represented in institution Snøhetta's winning design. The building includes many different ways to socially navigate and congregate in semiprivate spaces, including much clustered furniture and many semi-enclosed pods. Further, the materials of the building offer warm domesticity: colorful furniture and warm wood. It aims, like Halifax, to be a social haven, a "living room library." Both buildings are also exemplary in that they were constructed with environmental sustainability front of mind

Fig. 6.9 The Calgary Central Library includes many ways to socially navigate and congregate in semiprivate spaces including many semi-enclosed pods and optional digital use from a distance

(Georgijevic, 2016). This is especially true of the Halifax public library. It features a green roof, which uses rainwater to flush toilets (Taber, 2014).

6.5 To Take Away

With technology changing so quickly, it can seem that these changes are occurring so rapidly that it hardly feels like we can control our social reaction to innovation. Yet, design interventions can help us to leverage new possibilities and create a better future. Fostering social connection is one key way in which communication technologies will encourage sustainability: be it through Wi-Fi broadcasted from the back of a public school bus or in an open library collaboration space. Further, these new ways of connecting may help citizens to become environmentally engaged, for instance, through environmental assessment processes. These new facets of knowledge transfer also mean that knowledge is now global in nature and culture from around the world can be shared virtually, not just in situ. As to what the future holds, much of the direction of change will be determined by the companies that produce and our own technical infrastructure, so the meaning and weight of such design interventions must be balanced with the investment in the creation of accountability mechanisms for these corporations. The more accountable these corporations are, the more meaningful and inventive and powerful design interventions can be.

Questions for a Follow-Up Discussion

1. How was access to knowledge changed with the introduction of digital technologies? What future changes do you foresee?
2. To what extent has social media affected in-person interaction? Is social media eventually bound to replace people contacts?
3. How can the Internet be used to support remote learning?
4. How did digital technologies redefine the use of libraries and museums?
5. What additional cultural domains do you foresee will be affected by digital technologies?

References

Al-Arshani, S. (2020). School districts across the country are using school buses to deliver Wi-Fi to students who lack access. *Business Insider*. https://www.insider.com/wifi-buses-being-used-across-country-to-give-kids-internet-2020-3. Accessed April 26 2022.

Chen G.-M., & Sawyer, R. (2012). "The impact of social media on intercultural adaptation". *University of Rhode Island*. https://digitalcommons.uri.edu/com_facpubs/15/. Accessed April 28 2022.

Devlin, J. F., Yap, N. T., & Weir, R. (2005). Public participation in environmental assessment: Case studies on EA legislation and practice. *Canadian Journal of Development Studies/Revue canadienne d'études du développement, 26*(3), 487–500.

[EPD] Environmental Protection Department. (2015). "3D tools facilitate public participation in EIA process". http://www.epd.gov.hk/eia/englisremoh/press_release/3D.htm

Evans-Cowley, J., & Hollander, J. (2010). The new generation of public participation: Internet-based participation tools. *Planning Practice and Research, 25*, 397–408. https://doi.org/10.1080/02697459.2010.503432

Georgijevic, A. (2016). "Opening up the library". *The Globe and Mail*. https://www.theglobeandmail.com/life/home-and-garden/design/the-reinvention-of-libraries-from-public-spaces-to-livingrooms/article29333752/. Accessed April 26 2022.

Gow Hastings Architects. (2020) "Designing for higher education in a pandemic world". *Canadian Architect*. https://www.canadianarchitect.com/designing-for-higher-education-in-a-pandemic-world/. Accessed April 26 2022.

IAIA Session Discussion. (2016). "Innovative communication of project and impacts". *36th annual conference of the international association for impact assessment.*

Kohls, L. R. (2001). *Survival kit for overseas living*. Intercultural Press.

McClellan, S., Jimenez, J. A., & Koutitas, G. (Eds.). (2018). *Smart cities: Applications, technologies, standards, and driving factors* (p. 187). Springer.

Morgan, R. K. (2012). Environmental impact assessment: The state of the art. *Impact Assessment and Project Appraisal, 30*(1), 5–14.

Palfrey, J. (2015). *BiblioTech: Why libraries matter more than ever in the age of Google*. Basic Books.

Rose, K. [@kacierose4]. (n.d.). [TikTok profile]. *TikTok*. https://www.tiktok.com/@kacierose4?fbclid=IwAR0bf61sDc9aF3D_2zpq8dCvWlbY-bE6A9sne1LHrI5FC0iTeuK-t33lve4. Accessed April 28 2022.

Sinclair, A. J., Peirson-Smith, T. J., & Boerchers, M. (2017). Environmental assessments in the internet age: The role of e-governance and social media in creating platforms for meaningful participation. *Impact Assessment and Project Appraisal, 35*(2), 148–157.

Smith, B. (2021). "How TikTok reads your mind". *New York Times*. https://www.nytimes.com/2021/12/05/business/media/tiktok-algorithm.html. Accessed April 28 2022.

Taber, J. (2014). "Halifax looks forward to the opening of its very own library of the future". *The Globe and Mail*. https://www.theglobeandmail.com/news/national/halifax-looks-forward-to-the-opening-of-its-very-own-library-of-the-future/article20090514/. Accessed April 26 2022.

Townsend, A. M. (2014). *Smart cities: Big data, civic hackers, and the quest for a new utopia*. W.W. Norton & Company.

[TPL] Toronto Public Library. (n.d.). "Live & Online Programs". https://www.torontopubliclibrary.ca/programs-and-classes/. Accessed April 27 2022.

Turner-Lee, N. (2020a). "Bridging digital divides between schools and communities". *Brookings*. https://www.brookings.edu/research/bridging-digital-divides-between-schools-and-communities/. Accessed April 26 2022.

Turner-Lee, N. (2020b). "What the coronavirus reveals about the digital divide between schools and communities". *Brookings*. https://www.brookings.edu/blog/techtank/2020/03/17/what-the-coronavirus-reveals-about-the-digital-divide-between-schools-and-communities/. Accessed April 26 2022.

Wikipedia. (2022a). "Encyclopædia Britannica". https://en.wikipedia.org/wiki/Encyclop%C3%A6dia_Britannica. Accessed April 27 2022.

Wikipedia. (2022b). "Wikipedia". https://en.wikipedia.org/wiki/Wikipedia#Language_editions. Accessed April 27 2022.

Chapter 7
Well-Being, Public Engagement, and Healthcare

By 2050, an estimated 70% of the world's population is expected to live in cities. As a result, innovative ways of using technology among other measures will be needed to keep the growing urban population healthy and active. Virtual care, or the act of contacting a doctor through video calls, has grown in popularity in recent years as it allows for more flexible access to medical care for those living in remote communities and those with busy schedules or mobility issues. This chapter explores the synergy between public health, well-being, and digital means. It includes a discussion on the role technology has played in stymying the spread of infectious diseases, most recently experienced during the spread of the COVID-19 pandemic. The use of contact tracing and phone signals to predict outbreaks, for example, proves to be an effective strategy for mitigating the spread of the pandemic. Lastly, the relationship between social connection and health and how they can be fostered through technology will be discussed.

7.1 Cities for Well-Being

The contemporary city, as explored in other chapters of this book, is inundated with technology at nearly every turn. Townsend (2014) suggests that today there is a cacophony of billions of sensors tweeting from our pockets, the walls, and city sidewalks, reporting on minutia of every kind: vehicle locations, room temperatures, and seismic tremors, to name a few. It is nearly impossible to discuss each of the numerous roles that technology has in our lives. However, by focusing on the power that it has, to improve specific facets of our communities, we can imagine and create a world where technology helps build healthy and vibrant cities. Community and individual health are growing areas of technology development. By integrating certain new technologies into our cities and using existing ones, we can work to improve physical and mental health. The following sections explore devices and

A. Friedman, *The Sustainable Digital City*,
https://doi.org/10.1007/978-3-031-25488-8_7

applications (apps) that can be used by medical professionals, individuals, and cities to improve health.

7.1.1 *The Rise of Obesity and Engagement of People in Active Lifestyles*

In many nations, the average citizen is living an increasingly sedentary life. Most people spend many hours sitting at a desk or in a car, moving very little which increases the incidence of chronic illnesses. Cardiovascular diseases such as heart attacks and strokes are the leading cause of deaths worldwide (World Health Organization, 2017). Cardiovascular diseases are caused by inactivity, poor diet, and obesity. Sprawling, car-dependent neighborhoods and cities normalize and encourage a sedentary lifestyle. While medical advances have been vast, strategies to prevent illness should be prioritized, encouraging active lifestyles and working to educate the public on healthy habits.

Having an active lifestyle is naturally an individual decision. However, services and infrastructure can be put in place to encourage physical activity and active transit in cities. Well-maintained sidewalks and safe and efficient bike lanes are two major infrastructure investments that encourage active mobility (Fig. 7.1). Used for either recreation or active transit, bike lanes that prioritize the safety of cyclists and connect popular hubs throughout the city encourage more citizens to feel

Fig. 7.1 Well-maintained, safe, and efficient bike lanes like this one in Haarlem, the Netherlands, are investments that encourage active mobility and promote environmental sustainability

comfortable cycling. Divided bike lanes, advanced lights for cyclists, and colored paving help indicate the presence of a bike lane to motorists and protect cyclists. Proper bike lanes can encourage children and seniors to bike, knowing they are safe.

Public bike-sharing systems that were referred to in Chap. 3 were introduced in many cities around the world providing easy access to bikes across the city. They allow users to pick up a bike in one location and dock it at a station in a different part of the city. These bike-sharing programs mitigate the financial barrier to cycling and give users the freedom to use multiple forms of transit in one day. For example, a person could rent a bike to cycle to work and then walk or take public transit home. Renting a bike is relatively cheap and easy to do. New York's Citi Bike, for example, costs US$3.50 for a one-way 30-minute ride or US$15.00 for a day pass (City Bike, n.d.). Previously, to rent a shared bike, the user needed to have a credit card or a prepaid pass. However, in Montreal, the company has expanded its program to allow users to rent a bike with their public transit pass. Paying with a transit card gives users a 10% discount making the system more accessible and affordable (Bixi, n.d.). The Bixi app allows users to find docking stations around the city and see in real time if the stations have a bike to rent or spaces to dock a bike (Fig. 7.2). Users can also rent a bike through the app and plan their route (Bixi, n.d.).

China among other countries has begun implementing dockless public bike-sharing systems. In 2019, Chen et al. found that the dockless design of bike-sharing systems significantly improves users' experiences at the end of their bike trips. Individuals can instantly switch to a dockless shared bike without the responsibility of returning it back to a designated dock (Chen et al., 2020). Global positioning

Fig. 7.2 The Montreal Bixi bike-share app allows users to find docking stations around the city and see in real time if the stations have a bike to rent or spaces to dock a bike

system (GPS) trackers in the bikes allow users to find and rent bikes around the city where the program is implemented. Dockless bike-sharing systems are more flexible for users and often make the bike-sharing systems' integration with public transit even tighter than that of traditional public bikes, providing an efficient option for first−/last-mile trips (Chen et al., 2020). Increasing the flexibility and convenience of bike-sharing systems and public transit will encourage public participation. More on active and public transport can be found in Chap. 3.

While a sedentary lifestyle is a contributing factor to cardiovascular diseases, so is a poor diet. Access to a diversity of healthy, nutritious food is vital to good physical and mental health. However, access to healthy food is significantly determined by one's economic status and businesses in the community. When grocery stores do not offer affordable, good-quality, and delicious food, it is unlikely for people to choose those products. Urban agriculture in community gardens has been integrated in the planning of many cities. Rooftop gardens, community gardens, and hydroponic systems can be used to grow a variety of fresh, healthy food, in the same neighborhoods or cities as consumers (Figs. 7.3 and 7.4).

Fig. 7.3 At Benny Farm, a housing project in Montreal, Canada, the community gardens create a serene, contemplative atmosphere and support social interaction while providing valuable agricultural services nutritious food

Fig. 7.4 A variety of applications let people find fresh food near their home and exchange it with other backyard gardener's and recipes for inspiration

7.1.2 Technology and Medical Professionals: Virtual Care

Technologies are transforming the medical care that health practitioners are providing. With improved monitoring and interventions, life expectancy has increased significantly and is expected to continue to rise. For example, in 1970, the average life expectancy in Canada was 72 years; by 2020, it had risen to 82 years (O'Neill, 2019). Virtual care is the use of telephones or videoconferencing to connect patients with doctors and other health practitioners. The ability to consult with a physician from one's own home benefits many people, most notably those with limited mobility and people living in remote areas. Digital monitors positioned in a persons' bedside collect information from implants about their condition and send it to medical center for processing and monitoring. For example, one such system gathers information from a person's defibrillator to help cardiologists monitor and diagnose someone's health status (Fig. 7.5).

While in many nations the public health system is successful in eradicating many illnesses, there are gaps in services, long wait times, and challenges for people needing to access specialized care. Rural areas and remote communities struggle with a lack of access to specialized care and even basic care in some cases. Building medical facilities in and flying health practitioners to remote communities are costly and challenging. However, by connecting patients with physicians' online or over the

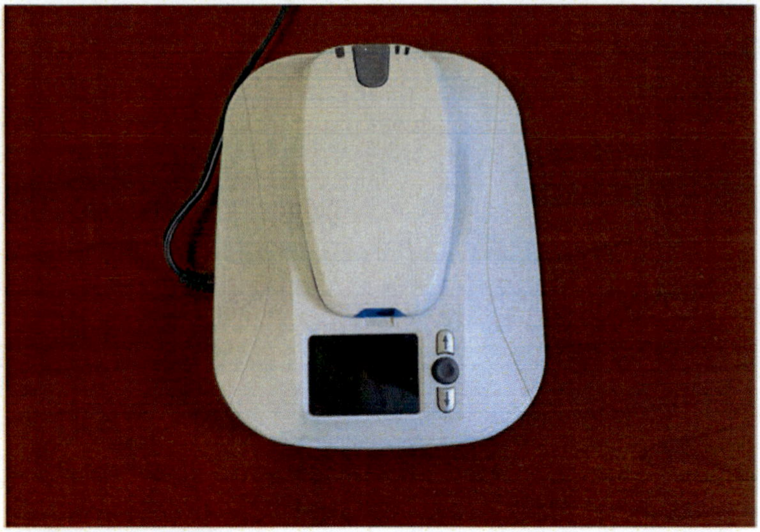

Fig. 7.5 Digital monitors positioned at a person's bedside collect information from implants about their user's condition and send it to a medical center for processing and monitoring. For example, one such system gathers information from a person's defibrillator to help cardiologists monitor and diagnose someone's health status

telephone, the healthcare provider can assess the patients' needs and determine an appropriate course of action. In many cases, a family doctor or general physician needs to refer a patient to specialized care, and virtual care can expedite the process. According to the Canadian Medical Association (2019), the World Health Organization (WHO) has projected that there will be a global shortage of some 18 million health workers by 2030 which will necessitate to make the existing system more efficient using technology.

Virtual care is a way to increase productivity of healthcare workers, allowing them to reach more patients in less time. The WHO has called on governments to assess their present and potential use of digital technologies for health, working to develop mandates for a global strategy to improve virtual care (Canadian Medical Association, 2019). In the 1970s, Dr. Maxwell House used telephone consultations to provide care to remote communities in Newfoundland, Canada, a revolutionary use of the technology at the time. In 2015, a Canadian telehealth report noted that there were 411,778 telehealth clinical sessions in 2014, representing just 0.15% of the 270.3 million billable services reported (Canadian Medical Association, 2019). The low rate of telehealth use is not due to lack of public interest; prior to 2020, 41% of Canadians wanted virtual visits with their provider, and only 4% of family doctors offered this option (Hardcastle & Ogbogu, 2020). While virtual care is a viable option in many circumstances, there are symptoms that are not amenable to virtual care: ear pain, cough, abdominal/gastrointestinal symptoms, musculoskeletal issues, most neurological symptoms, and congestive heart failure (Hardcastle & Ogbogu, 2020).

The spread of COVID-19, and subsequent wide lockdowns, rapidly accelerated the adoption of virtual care. Since in many nations healthcare provision is overseen by local governments, different models for virtual care are used. In Canada, for example, there are two models for virtual care. The first allows patients to contact their family doctor or other physicians who they have seen before, while the second is a virtual equivalent of a walk-in clinic (Hardcastle & Ogbogu, 2020). Accessing virtual care does require some technology literacy which can be a hurdle for some seeking care; however, through greater investment in developing telehealth technologies, the platforms for accessing care can be improved.

Incorporating technology into healthcare is not limited to phone calls or videoconferences—booking appointments, ordering prescriptions, or requesting personal medical records can be done online. A survey found that 71% of Canadians would like to be able to book appointments electronically, but just 9% of family physicians presently offer this option. There is a clear gap between the electronic access that patients would like to have and what physicians are presently offering (Canadian Medical Association, 2019). Virtual care is not limited to services provided by general physicians; professionals such as physiotherapists are increasingly including tele-rehabilitation in their practices. Tele-rehabilitation applies to all aspects of patient care including patient interview, physical assessment and diagnosis, treatment, maintenance activities, consultation, education, and training (College of Physiotherapists of Ontario, n.d.). Depending on the clinicians' practice, telerehabilitation can incorporate videoconferencing, email, apps, web-based communication, and wearable technology (College of Physiotherapists of Ontario, n.d.). Videoconferencing applications such as Zoom can be complicated to use for practitioners and patients; the platform requires the app to be downloaded on a device and can require updates, delaying appointments. Entering the chat room can be difficult at times, and when providing virtual care, time is important. Healthcare providers and patients alike do not want to spend precious time resolving technical issues. To mitigate this challenge, platforms such as Doxy.me have been created. It is a secure telemedicine videoconferencing platform that allows healthcare providers to create a waiting room or patient queue (Doxy.me, n.d.). The platform is accessed simply by going to the URL provided by the clinician (doxy.me/YourDrsName). Simple technology advances like this greatly improve the clinician and patient experience and healthcare provision.

Virtual care has also been extended to mental health services. Therapy provided online or by telephone is increasingly common and has made mental health services more accessible. There can be many barriers to accessing therapy such as a busy work schedule, domestic responsibilities, and challenges commuting. Researching the efficacy of online versus in-person care, Jonathan Cook and Carol Doyle found that no significant differences in the level of working alliance were found within the online therapy sample with respect to the modality of communication, client presenting problem, or therapist (Cook & Doyle, 2002). While some patients might prefer in-person therapy, online mental health services significantly improve its accessibility. Platforms such as BetterHelp operate exclusively online to provide care from licensed, trained, experienced, and accredited psychologists (PhD/PsyD), marriage and family therapists (LMFT), clinical social workers (LCSW/LMSW), and board licensed professional counselors (LPC) (BetterHelp, n.d.). The platform allows users to message, live message, speak over the phone, and videoconference with a therapist depending on the user's preference. There are also a growing number of talk and text mental health services available around the world. Programs such as Crisis Services Canada, a website that directs users to helplines in their area and is available to everyone in Canada. Kids Help Phone is available to individuals between 5 and 29 years old, and Hope for Wellness Helpline crisis intervention for Indigenous people in Canada provides free telephone and online support (Government of Canada, 2020). Many of these services are offered 24 hours a day, every day. While there is still a large need for affordable and convenient access to regular therapy, crisis services that are available at any time significantly improve the care available to people in need of acute care. Advances in medications to treat mental illness have been crucial to improving the quality of life and well-being of millions of people around the world. However, research generally shows that psychotherapy is more effective than medications and that adding medications does not significantly improve outcomes from psychotherapy alone (American Psychological Association, n.d.). Psychotherapy must be provided alongside medication if significant and long-lasting wellness is to be achieved.

Support groups for those struggling with addiction, mental illness, and physical illness as well as for family members of people experiencing suffering are increasingly available online. In an age with heightened social isolation, online support groups and therapy can provide support to those who need it and a sense of connectedness, that they are not alone. Support groups offer people a chance to share their experiences, learn coping techniques, and be involved with a community.

7.2 Engaging Citizens

Technology has been used by many to gather people with similar goals and ideas. Social media platforms give people the power to organize and speak about issues that are important to them. Social media also allows us to share personal updates with friends, family, and the broader public and find people opinions and display them for all to see (Fig. 7.6). Writers and content creators have been creating information about exercise, active lifestyles, and healthy eating for decades. Information is freely available online about how to improve one's health and well-being. But engagement and encouraging people to make a significant change toward the better can be a challenge. This section looks at the policies and technologies that can be used to engage citizens.

Fig. 7.6 Specialized applications allow communities to find people's opinions and display the outcome for all to see

7.2.1 *Devices to Monitor Health and Motivate*

The Internet is increasingly having a significant influence on people's decisions about their health due to the vast amount of information available online. However, not all the available information is reliable, accurate, and safe. The COVID-19 pandemic made this especially clear, with new "miracle cures" being presented online and in the media as safe and effective despite the cures being either highly dangerous or simply ineffective. Access to reliable, trusted, and safe information is increasingly important in a time when patients not only access online health information but also make changes based on this information, with only about half of all patients sharing this information with their physicians (Liszka et al., 2006).

In an effort to combat the mass of contradictory information about COVID-19, social media platforms such as Twitter, Instagram, and Facebook have added pop-ups and stickers on all posts that mention COVID-19, directing users to information published by the WHO and local health ministries (Instagram, 2020). The need for accurate and reliable information from national and international health ministries is needed far beyond the scope of COVID-19. Platforms that provide reliable health guidance to citizens and encourage healthy lifestyle habits are needed to improve community health. Wearable technologies are becoming more popular as a method of monitoring different conditions and can be used by the individual or health practitioner to develop a better understanding.

Wearable technologies are electronic devices that consumers can wear, like Fitbits and smartwatches, and are designed to collect the data of users' personal health and exercise (Phaneuf, 2021). The monitoring devices can be used to track vital signs and health. The driving thought behind wearable technology is the notion that by recording and reporting information about behaviors such as physical activity or sleep patterns, these devices can educate and motivate individuals toward better habits and improved health (Asch et al., 2015). These devices are giving their users the ability to access notifications from their smartphone. Depending on the complexity of the device, they monitor the user's heart rate, count steps, blood oxygen saturation, physical activity, and sleep patterns. Some devices can send user's health information to a doctor or other healthcare professional in real time (Phaneuf, 2021). In some cases, the devices can alert those who are experiencing atrial fibrillation and notify them that they should access medical treatment (Phaneuf, 2021). Advanced wearable medical devices include but are not limited to electrocardiogram (ECG) monitors, blood pressure monitors, and biosensors. The main target of these devices is to track the users' vital signs and notify them when irregularities occur that could be potentially life-threatening before physical symptoms become noticeable. Utilization of the digitalized patient health monitoring can facilitate diagnosis, treatment plans, and physician–patient interaction and accelerate the progress of medical research, education, and training (Shinbane & Saxon, 2016).

Wearable devices are also used by many people to track their personal fitness and health. Smartwatches can be used to track walking, running, cycling, and other physical activities. Using running as an example, the device can track the user's route, pace, time, incline of the route, and speed. By collecting this data over a period of time, the affiliated app can notify the user of improvements in performance and goals achieved. Some apps recommend routes to follow and allow the user to share their statistics on social media. Other apps help users track their sleep and nutrition. By logging daily habits, users can see aggregated information over time and take steps to improve in areas they want to. In their paper *Opportunities and Challenges for Smartphone Applications in Supporting Health Behavior Change: Qualitative Study*, Laura Dennison et al. (2013) suggest that accuracy and legitimacy, security, effort required, and immediate effects on mood emerged as important influences on app usage. The ability to record and track behavior and goals and the ability to acquire advice and information "on the go" were valued. Social media features were not seen as important and were considered off-putting; however, participants in the study did express interest in technology that supports "health-related behavior change" (Dennison et al., 2013). The challenge with wearable technologies and health apps is to bridge the gap between awareness and action. Once personalized health information is made available, the user then must decide to change their behaviors toward healthier habits.

7.2.2 Engaging Seniors and Aging in Place

Many nations have a large elderly population segment. For example, Canada has nearly 17 million people aged 45 years and over (Varrella, 2021) (Fig. 7.7). As the population ages, it is important to provide services and programs that support health and well-being. While medical interventions have become increasingly successful,

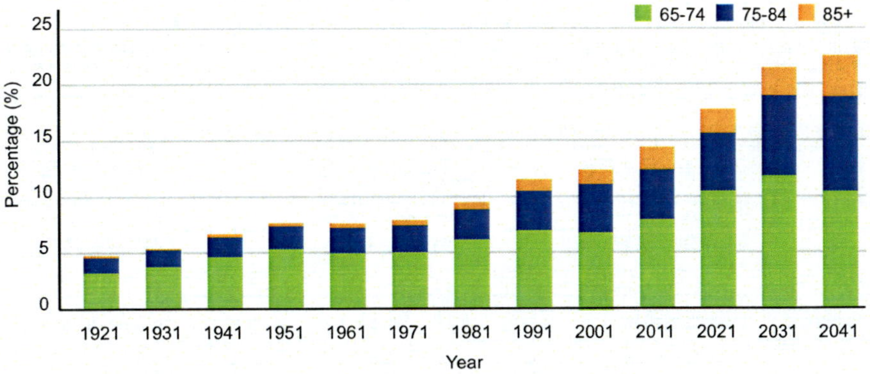

Fig. 7.7 Canadian seniors by age groups as percentage of the total population, 1921–2041

disease and illness prevention is important to preserve the quality of life and happiness of individuals and the broader community. While all the technological interventions explored above are available to the whole population, there are different needs for an aging population and different interventions that can be done.

While learning new technologies can be difficult, there is a growing need for technology literacy. There is an abundance of information and services available online, but what happens when people have difficulty accessing this information? More advanced and complicated technology is not necessarily better. Simplicity and accessibility are important to preserve when developing technologies from website design to wearable devices. Online services targeted toward seniors should be easy to use and designed to reduce user errors. The Advanced Communications Law and Policy Institute (ACLP) found that 53% of seniors use the Internet to get information about healthcare or medical issues and 75% use it to communicate with friends and family (Schaub, 2014). With the growing adoption of smart appliances, caregivers can use online platforms to provide better care for seniors. They may control appliances, home security, and lighting in a senior's home even when they, the caregiver, are not in the home. Appliances and systems integrated in the home can be controlled or monitored on apps from outside the home, allowing caregivers to ensure the safety of their patients.

Providing service and fostering relationships among elder community members is important for their health and well-being. For example, online fitness classes geared toward an aging population safe and easy to follow at home, with limited equipment, can help improve the physical health of seniors. Virtual retirement communities are an example of the integration of technology into senior care and the understanding that many individuals want to stay independent, living in their own home, but need or want some assistance. The virtual community is staffed by a team who organizes a multitude of services and programs for members. These may include transportation, assistance with medical appointments, grocery shopping, organized group outings, and aid with applying for benefits. One co-founder of a virtual retirement community explained that having a virtual village means staying in touch with others and connecting to services in a way that wasn't possible 10, 20, or 30 years ago (Cloud, 2014). These communities are striving to support successful aging in place by providing easy access to services and connections that many people need, sharing hobbies of interests, providing new learning opportunities, financial advice, nutrition tips, exercise videos, mental health support, and games to strengthen cognition such as sudoku and crosswords. The communities are very affordable, costing between CAN$125 and CAN$500 per year (Gerace, 2013). Virtual retirement communities are excellent examples of incorporating technologies to engage seniors and improve their quality of life.

In an insightful article titled, *Social Interaction Is Critical for Mental and Physical Health*, Jane Brody explores the need for a lack of technology for better health. Social isolation is something that many people struggle with. While social isolation can be a result of many factors such as age, job loss, living away from family, mental illness, and limited mobility, it has serious consequences. Social isolation is on a par with high blood pressure, obesity, lack of exercise, or smoking as a risk factor for illness and early death. According to Brody (2017), those with close social ties and unhealthful lifestyles such as smoking, obesity, and lack of exercise actually lived longer than those with poor social ties but more healthful living habits. While disease prevention should include efforts to increase exercise, healthy eating, and not smoking, fostering social connections can be just as lifesaving. Brody presents an argument for actually removing technology in order to create spaces that bring people together. Some cafes in New York, Toronto, and Montreal have adopted this strategy by not providing Wi-Fi to customers or restricting the use of laptops in certain sections of the space. While it might be unfamiliar or uncomfortable for some people, this sort of informal gathering offers people the opportunity to connect with other community members, creating friendships or simply enjoying a conversation together. Brody (2017) suggests that people who are chronically lacking in social contacts are more likely to experience elevated levels of stress and inflammation. These, in turn, can undermine the well-being of nearly every bodily system, including the brain. The Centers for Disease Control and Prevention (CDC) reported in 2020 that one-third of adults aged 45 and older feel lonely and nearly one-fourth of adults aged 65 and older are considered to be socially isolated (CDC, 2020). Fostering social connections and having the support of friends and family is integral to our health and longevity. By building communities that provide care and support for all members of the population, we will create healthier and happier lives.

7.3 Technology for Everyone

By integrating technology into medical professions and communities, public well-being and health can be improved. Public bike-sharing systems have been integrated into cities and communities around the world, providing people with a reliable and affordable opportunity to exercise and participate in active transit. Access to locally grown, nutritious food encourages people to eat healthier and decreases the community's carbon footprint. Virtual care makes health services more accessible, and finding medical advice and information online empowers individuals to better understand their health and make lifestyle changes. Virtual mental health services have made care more accessible and affordable. Crisis support and long-term talk therapy services protect people when they are experiencing difficulties. Wearable technologies allow users to monitor their physical health and exercise achievements as well as notify the wearer in the event of potentially dangerous cardiac irregularities. Technology can be used to engage and educate seniors, providing health and financial information as well as fostering social connections. Programs such as virtual retirement communities are working to help seniors age in place while recognizing the importance of community and access to services. This chapter has explored some of the ways technology can be incorporated into our cities to improve the health and well-being of its residents, strengthening the community.

Questions for a Follow-Up Discussion

1. What are the risks that inactivity and obesity pose to personal health?
2. What is virtual care, and how is it changing the delivery of public healthcare?
3. How can digital means engage citizens in communal activities?
4. What are wearable technologies, and how can they contribute to monitoring people health?
5. How can digital technologies support aging in place?

References

American Psychological Association. (n.d.). "How Do I Choose Between Medication and Therapy?". from https://www.apa.org/ptsd-guideline/patients-and-families/medication-or-therapy#:~:text=Research%20generally%20shows%20that%20psychotherapy,improve%20outcomes%20from%20psychotherapy%20alone. Accessed April 14 2022.

Asch, D. A., Patel, M. S., & Volpp, K. G. (2015). Wearable devices as facilitators, not drivers, of health behavior change. *JAMA, 313*(5), 459–460.

BetterHelp. (n.d.). "About". https://www.betterhelp.com/about/. Accessed April 14 2022.

Bixi. (n.d.). "Pricing". https://www.bixi.com/en/pricing. Accessed April 14 2022.

Brody, J. (2017). "Social interaction is critical for mental and physical health". https://www.nytimes.com/2017/06/12/well/live/having-friends-is-good-for-you.html. Accessed April 26 2022.

Canadian Medical Association. (2019). "Virtual care in Canada: Discussion paper". CMA Health Summit. https://www.cma.ca/sites/default/files/pdf/News/Virtual_Care_discussionpaper_v2EN.pdf. Accessed April 15 2022.

Center for Disease Control (CDC). (2020). "Loneliness and social isolation linked to serious health conditions". https://www.cdc.gov/aging/publications/features/lonely-older-adults.html. Accessed April 15 2022.

Chen, Z., van Lierop, D., & Ettema, D. (2020). Dockless bike-sharing systems: What are the implications? *Transport Reviews, 40*(3), 333–353.

City Bike. (n.d.). "Pricing". https://www.citibikenyc.com/pricing. Accessed April 14 2022.

Cloud, R. (2014). "A virtual retirement community". https://retirementandgoodliving.com/a--virtual-retirement-community/. Accessed April 14 2022.

College of Physiotherapists of Ontario. (n.d.). "Virtual Practice". https://www.collegept.org/registrants/virtual-practice-in-physiotherapy. Accessed April 14 2022.

Cook, J. E., & Doyle, C. (2002). Working alliance in online therapy as compared to face-to-face therapy: Preliminary results. *Cyberpsychology & Behavior, 5*(2), 95–105.

Dennison, L., Morrison, L., Conway, G., & Yardley, L. (2013). Opportunities and challenges for smartphone applications in supporting health behavior change: Qualitative study. *Journal of Medical Internet Research, 15*(4), e86.

Doxy.me. (n.d.). "For patients". https://doxy.me/en/patients/. Accessed April 14 2022.

Gerace, A. (2013). Virtual retirement communities see rising membership, Interest. Senior Housing News. https://seniorhousingnews.com/2013/07/22/virtual-retirement-communities-see-rising-membership-interest/. Accessed April 14 2022.

Government of Canada. (2020). "Mental health support: Get help". https://www.canada.ca/en/public-health/services/mental-health-services/mental-health-get-help.html#:~:text=Call%201%2D800%2D668%2D,anonymous%20care%20from%20professional%20counsellors. Accessed April 14 2022.

Hardcastle, L., & Ogbogu, U. (2020). Virtual care: Enhancing access or harming care?. In *Healthcare management forum* (Vol. 33, No. 6, pp. 288–292). SAGE.

Instagram. (2020). "Keeping People Informed, Safe, and Supported on Instagram". from https://about.instagram.com/blog/announcements/coronavirus-keeping-people-safe-informed-and-supported-on-instagram. Accessed April 14 2022.

Liszka, H. A., Steyer, T. E., & Hueston, W. J. (2006). Virtual medical care: How are our patients using online health information? *Journal of Community Health, 31*(5), 368–378.

O'Neill, A. (2019). "Life expectancy in Canada, 1800–2020". Statista. https://www.statista.com/statistics/1041135/life-expectancy-canada-all-time/. Accessed April 14 2022.

Phaneuf, A. (2021). "Latest trends in medical monitoring devices and wearable health technology". Insider. https://www.businessinsider.com/wearable-technology-healthcare-medical-devices. Accessed April 14 2022.

Schaub, H. (2014). "Why senior citizens use the internet". https://www.brookings.edu/blog/techtank/2014/04/11/why-senior-citizens-use-the-internet/. Accessed April 14 2022.

Shinbane, J. S., & Saxon, L. A. (2016). Digital monitoring and care: Virtual medicine. *Trends in Cardiovascular Medicine, 26*(8), 722–730.

Townsend, A. M. (2014). *Smart cities: Big data, civic hackers, and the quest for a new utopia.* W.W. Norton & Company.

Varrella, S. (2021). "Canada—resident population, by gender and age group 2020". Statista. https://www.statista.com/statistics/444858/canada-resident-population-by-gender-and-age-group/#:~:text=The%20average%20age%20of%20the,males%20and%20females%2C%20are%20single. Accessed April 14 2022.

World Health Organization. (2017). "Cardiovascular diseases (CVDs)". https://www.who.int/news-room/fact-sheets/detail/cardiovascular-diseases-(cvds). Accessed April 14 2022.

Chapter 8
Home Automation

What once could be considered an idea out of science fiction, home automation holds the potential to transform the efficiency and functioning of the domestic environment. Parallel to the integration of digital means in cities, a similar mindset and practice can be applied to the home. While full-digital residential systems are costly to some, they can potentially reduce utility bills for decades once implemented. This chapter reviews the five most common contemporary axes of home automation technology to include irrigation management, automated lighting, smart power adapters, and intelligent thermostats.

8.1 The Rise of the Digital Home

The integration of smart technologies into the residential environment has increased in recent decades and, along with market demand, will likely continue to do so. The concept of the digital or smart home is not new. It has been discussed and explored in the media over the past century, depicting numerous possibilities and benefits that a future filled with such gadgets could bring to individuals and families. Innovations were theorized with the belief that technological innovations could reduce energy costs, time spent doing chores, and water usage in the home. In a 1939 article in *Popular Mechanical Magazine*, possible technological advances were speculated to soon arrive. Some would include control centers dispersed throughout a home, microwave ovens, small-scale hydroponic farming, DVRs, and alternative home temperature conditioning strategies, to name a few (Bucher, 1939).

An obvious and culturally significant reference for these beliefs was the popular cartoon, The Jetsons, first released in 1962. The series follows the everyday antics of a family living a century in the future and positions the home in the center of both the world view and as the site of technological innovation. Different inventions are scattered throughout the episodes as the characters interact with them, like a smart

A. Friedman, *The Sustainable Digital City*,
https://doi.org/10.1007/978-3-031-25488-8_8

fridge that ejects instant meals as if it were a vending machine and a robotic maid that takes care of petty chores around the house (Sessums, 2020). These technological devices are introduced as ways to increase the efficiency of tasks in the home and provide the family with more free time, pointing to the interest at the time to reduce domestic work. Another popular film was Disney's live-action movie titled *Smart House* released in 1999 and directed by LeVar Burton. The movie adopts the notion of the home as characterized figure that is connected to almost all living aspects, having their own personality, voice, and ability to interact autonomously with its occupants. The voice, known as PAT, is in some ways a successful prediction of voice-commanded smart home technologies (SHTs) that would appear in the decades that followed (Bhardwaj, 2018). The latter example demonstrates ideas that improve the living comfort of the occupants. In recent years, however, the perspective has adapted to the unprecedented context of the climate emergency our planet is facing and shifted focus on how SHT can increase a dwelling's sustainability.

8.1.1 The Digital Effect on Home Design and Use

A smart home is characterized by technological solutions that enable the automation of electronic, electrical, and technology-based tasks within the home, combining hardware and software technologies. The topic is closely related to and can be expanded even further within the concept of the Internet of things (IoT). The IoT consists of everyday physical objects being connected to the Internet and able to identify themselves to other devices. The significance of this framework is derived from the ability of an object to represent itself digitally and become part of something greater than itself; the object is no longer understood solely in its relation to the user but in its relationship with surrounding objects and metadata. When many objects act in unison, they are referred to as having *ambient intelligence*. Smart technologies can and have integrated these concepts to significantly affect the organization and purpose/use of the home in multiple ways over the past century and are integral to improving automation in housing design at a community scale. A 2018 Forbes article estimates that by the end of that year, 45 million smart home devices would have been installed in households across the United States (Caccavale, 2018). This is due to a multitude of factors, including most importantly a rise in energy costs attributed to the daily functions of the home as well as a decrease in the cost of these technologies (Caccavale, 2018). The home is no longer solely a place of living. Recent trends demonstrate the increasing importance of flexibility in the home as it has become desirable, especially over the course of the COVID-19 pandemic, to work and perform other tasks from home without needing to travel elsewhere. On top of this, self-employment is rising steadily. Out of 57 million freelancers in the United States, 50% of them were millennials, while in the United Kingdom, the figure jumped from 3.3 million in 2001 to 4.8 million in 2017, equating to about 15% of the labor force in 2017 (Design Build Network, 2018).

8.2 Smart Home Technologies (SHTs) and Sustainable Design

SHTs are integral to achieving efficient automation in a dwelling. According to a paper by Furszyfer Del Rioa et al. (2021), a device must check off certain boxes to be considered "smart." These criteria are the following: they should be digitally connected, provide enhanced control options to the user(s), allow for automated processes, and have the capability to learn. With the number of devices connected to the Internet projected to reach 21 billion by 2025, and 880 million devices sold in 2019 alone, these tools have the potential to introduce sustainable practices from the scale of the home to that of the city (2021) (Fig. 8.1).

As outlined in Chap. 2, sustainability can be categorized into four main pillars, each having the potential and ability to effectively integrate ideas of the smart home into its theoretical framework. The first, *economic*, relates automation to home functions cost efficiency. For example, heating could be regulated based on outdoor temperatures and automatically adjusted to prevent unnecessary heating. *Social* sustainability is the next pillar and can be used to contextualize how technology can aid

Fig. 8.1 Digital features that are presently available for occupants of smart homes

Fig. 8.2 Home devices like Amazon's Echo or Dot are solutions that can bridge the communication between elderly people and their relatives

vulnerable populations to retain autonomy in their place of living. Communication between the elderly and their relatives and/or friends can become challenging, especially during the COVID-19 pandemic that has affected the world, and it is not always possible to meet in person. Home devices like Google Home, Amazon's Echo Dot, or Alexa are available solutions that can bridge these gaps in an accessible way (Fig. 8.2). *Environmentally*, technological tools can record power and water usage and terminate their use when a specific task has been completed. For example, automated irrigation management systems can greatly reduce the amount of water wasted when watering plants. Smart devices equally impact *cultural* sustainability by expanding what is possible and opening a new frontier of how we interact with the built environment and other people. Home automation solves problems at a microscale, but if integrated into society, the technology could have larger and more impactful ramifications. Communities centered around automation would not only make mundane tasks obsolete, but if guided by sustainability then they would encourage the development of more sustainable technology.

8.2.1 Hubs and Controllers

Hubs and controllers are one of the five broad categories of smart devices mentioned above and contribute primarily to building and maintaining social sustainability. They can help older populations retain some of their autonomy in the home and contribute to the concept of independent living. As mentioned above, the integration of these smart devices in seniors' homes can greatly improve their health, quality of life, and safety. According to Sociotio (2020), aging in place is the ability to live in one's own home and community safely, independently, and comfortably, regardless of age, income, or ability level, meaning that housing should accommodate and be adapted to all people as to avoid displacement. Studies and surveys have demonstrated that a sizable portion of the population in the United States, over 111 million, will be over the age of 65 by 2050 and that a vast majority prefer to live in their own home as opposed to a senior's residence (Sociotio, 2020). Smart hubs and controllers can be utilized to increase and promote aging in place—not to mention that disabled individuals of any age and those with reduced mobility or autonomy can equally benefit from these technologies (Sociotio, 2020).

The health of seniors and at-risk ill people can be monitored effectively from the home with the help of smart devices. For example, a health monitoring system could be set up, allowing for nurses and/or primary care practitioners to receive health reports directly from their home, as well as send them back from their care facility. Similarly, a smart fridge might aid in ensuring that they are maintaining a healthy diet depending on their individual needs by assessing what is currently stocked in the fridge and subsequently generating list of what is missing and will need to be bought so that certain items are not forgotten (Constellation, 2018). Furthermore, voice assistants can regularly remind people to take essential medications, and robot vacuums, such as the Roomba by iRobot, can aid in keeping the house clean, especially important for those with reduced mobility. Memory loss also comes with age. Speakers with built-in assistants can order cars on rideshare applications through voice commands, while smart assistants can easily make calls using a person's contact list, making it easier for seniors to make calls and keep in touch with family, caretakers, and friends. All these services essentially eliminate—or at least greatly reduce—the premature relegation of the elderly to assisted living facilities and allow for increased autonomy. As a result, the diversity of ages in a neighborhood can remain intact for an extended time and enrich a community. Doubly, caretakers can feel less stressed about the condition of their loved ones knowing that they are being monitored for their health and safety and that contact is facilitated.

8.2.2 Smart Water Management (SWM) and Irrigation

Water scarcity is poised to become one of the major imminent global environmental crises in the near future. The United Nations (UN) suggests that water use has been growing globally at more than twice the rate of population increase in the last century, and an increasing number of regions are reaching the limit at which water services can be sustainably delivered, especially in arid areas (UN Water, 2021). Reports by UN Water also indicate that municipalities account for 16% of global water withdrawals (Fig. 8.3). The introduction of smart technologies for the management and use of water supplies stands to increase the sustainability of homes and communities.

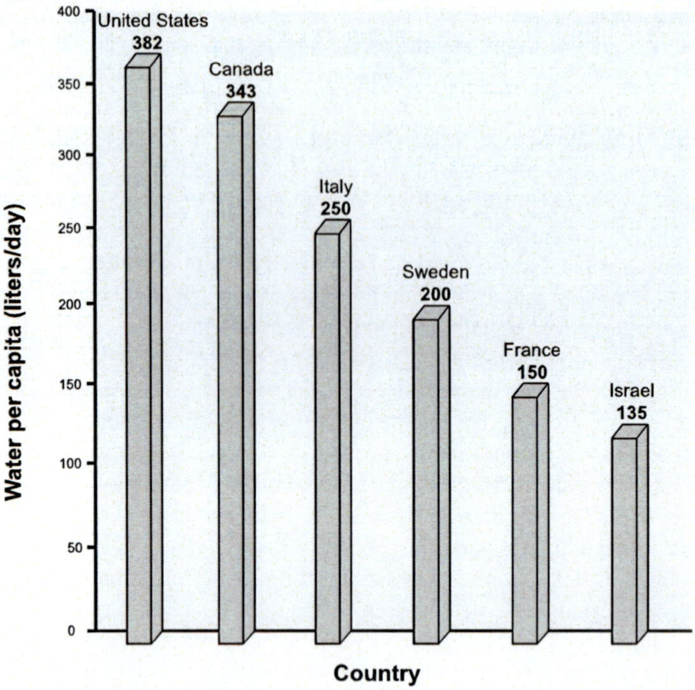

Fig. 8.3 Water consumption per capita of several nations

SWM systems are one way of tackling this challenge at a multitude of scales. The advantages of integrating SWM systems is that they can reduce excessive water use, improve the efficiency of infrastructure, control leakage, and monitor consumption at the household scale, among other benefits (Digiteum Team, 2021). Municipal water use can be monitored in real time with the help of sensors. Reports by the sensors can indicate where problems lie in terms of unsustainable practices and what can be improved whether infrastructural or habitual to consumers on their cellular devices and laptops, thereby increasing the accessibility of this information and the autonomy of individuals to react accordingly (Li et al., 2020; Digiteum Team, 2021). This works hand in hand with smart irrigation systems which will water one's yard depending on the amount of moisture in the soil, humidity, and weather (Digiteum Team, 2021). There already exist several smart sprinkler controllers, for example, that will factor in and adjust to the outdoor conditions to reduce water use and that are compatible with varying home controllers and hubs (Tuohy, 2021). In a trial run by ITA, a smart home company based in Melbourne, Australia, in which 60 households were provided with irrigation controllers, they found that water use was reduced by 15% or 95,000 liters (25, 096 gallons)—equivalent to about two backyard swimming pools per year (2021). Applying these technologies at the scale of entire communities can significantly reduce water consumption and improve the environmental sustainability of households.

8.2.3 *Automated Lighting, Blind Control, and Smart Power Adapters*

Excessive and inefficient energy use is another area of concern that can be improved to increase the sustainability of homes. Although the typical lightbulb has evolved over recent years to reduce energy waste, automation of artificial lighting in the home can be equally integrated into automated systems to have a greater impact in reducing energy use. In addition, blind control technology can be employed to automatically pull up or down in response to the exterior conditions or on a schedule (Fig. 8.4). A report that studied the effects of blind control on lighting demand and heating and cooling requirements found that the use of automated blinds reduced lighting demand by a considerable amount (Chung et al., 2019). Consequently, cooling demands, which will become more important as global temperatures rise, could also be reduced by a significant margin (Chung et al., 2019). By employing automated lighting that works in congruence with blind control systems in the home, excess energy use can be avoided and benefit environmental sustainability.

Smart plugs are an alternative to regular plugs that seek to reduce energy demand in the house with added convenience to users, contributing to the economic and environmental pillars of sustainability. Other devices, whether smart devices or not, can be plugged into the smart plugs. Doing so provides flexibility to users, allowing

1	Heating	4	Blinds
2	TV / Wi-fi	5	Camera
3	Living room lights	6	Dining room lights

Fig. 8.4 A living–dining room space with remote-controlled Wi-Fi features

them to access their devices from anywhere in the home using their smartphone or laptop and without having to go touch them (Fig. 8.5). Certain devices, blinds or lights, for example, can be put on a schedule when connected to smart plugs via the user's phone and a laptop (Louis et al., 2015). Additionally, the energy consumed by each appliance plugged into the smart plugs can be monitored and synthesized into a report for homeowners, providing them with information on what can be improved in terms of energy efficiency—information which is easily accessible by phones and other electronics (Bradford, 2021) (Fig. 8.6).

8.2.4 Intelligent Thermostats and Thermal Comfort

Heating and cooling machines also contribute to the climate crisis. As more heating and cooling is required to regulate interior temperatures due to climate change, more energy is consumed, and greenhouse gases are released into the atmosphere, accelerating climate change and the cycle. About 100 million homes in the United States have mechanical air-conditioning according to the US Energy Information Administration (EIA) (2011); it is estimated that heating and cooling accounts for 50% of the energy used in US households (Foster, 2021). A significant portion of homes, specifically 20% of those built in the 1980s, employ air-conditioning systems that are over 20 years old and highly inefficient (US EIA, 2011). Many, if not most, buildings in the world lack autonomy and the ability to react to the climactic conditions of their respective contexts. Instead, tried-and-true engineering

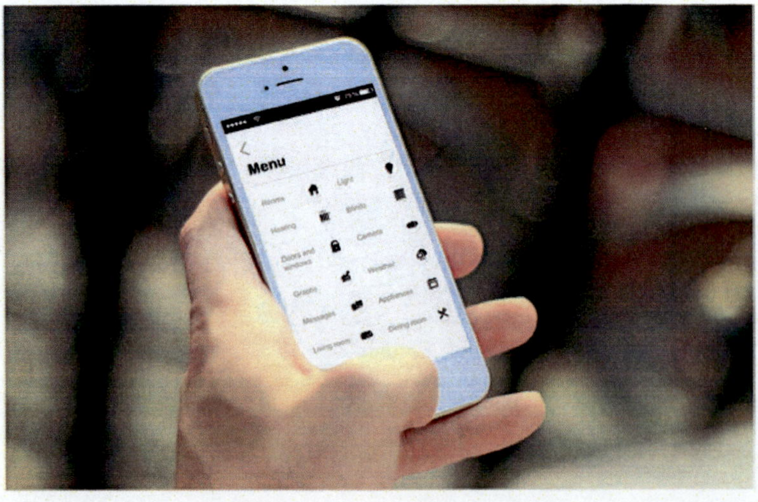

Fig. 8.5 Users of home automation's features can access devices from a distance using a smartphone

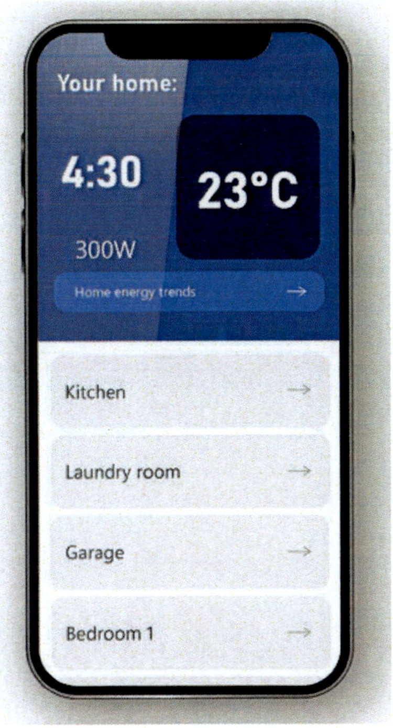

Fig. 8.6 The energy consumed by each appliance plugged into the smart plugs can be easily accessible for reuse by phones

principles and rules of thumb are applied to most buildings due to their economic benefits, convenience, and ability to control.

If society wishes to combat its dependence on these appliances, the way buildings are designed needs to change to mitigate the effects of climate change while keeping people as comfortable as possible. Recent studies and research on alternative modes of ventilation have been conducted and demonstrate that mixed-mode ventilation schemes are not only possible but favorable. Mixed-mode ventilation refers to a ventilation strategy which makes use of the exterior air to stabilize indoor temperatures while maintaining the thermal comfort of individuals inside alongside mechanical heating, ventilation, and air-conditioning (HVAC) systems typically used in contemporary buildings (de Dear & Brager, 2021). These systems are connected to decrease energy use and have different smart devices working together and off each other's input into the system. For example, operable windows in a home could be programmed to open when the exterior temperature is mild, which in turn signals the HVAC system to temporarily shut down, therefore saving on energy and the associated environmental damage and financial costs. Systems can also be installed to monitor and manage energy use in real time associated with heating and cooling and provide consumers with this information (Lenz et al., 2011). Intelligent thermostats can be used to automatically regulate indoor temperatures and reduce energy consumed from heating and cooling (Fig. 8.7). According

Fig. 8.7 Intelligent thermostats can be used to regulate indoor temperatures automatically reducing energy consumed by heating and cooling

to Google Nest, the Nest Learning Thermostat reduced heating and cooling costs by 10–12 and 15%, respectively, for households in the United States (Bogart, 2016).

8.2.5 Household Energy Conservation Games

A large aspect of the reduction of household energy consumption is the mindset of the homeowner and other dwellers. Active thought must be put into installing and maintaining energy-saving smart home devices; those motivated by environmental or monetary factors might find these tasks the easiest. Researchers Fijhneer et al. (2021) from Utrecht University in the Netherlands, however, have found another way to get people on board and motivate them into engaging with energy conservation. They have developed a competitive household energy conservation game that reduces energy consumption in the home by 8%. Real-life home energy use is obtained digitally from a smart energy meter and incorporated into the game. It has quiz features educating players on energy conservation methods, avatars for each household member, competition aspects, and different missions with the goal of reduction of energy consumption. This process of gamification—incorporating game features into education—is a valuable motivational tool for making nongame events more engaging. The competition aspect of the game especially in a family and/or household dynamics furthers motivation for success. The conclusive findings were that persuasive educational games are effective in helping people lower energy use in both the short and long term (2021). This game and potentially others like it are useful tools to convince those who are not economically or environmentally motivated to reduce energy consumption to, in fact, do so.

8.3 Personal Privacy and Accessibility

There are a couple reasons why an individual might hesitate to buy and install smart devices in their homes despite the benefits previously listed. These concerns might relate to issues of personal privacy and data collection, elevated upfront costs, and confusing interface and/or programming, to name a few. Many feel a sense of insecurity and are concerned with their privacy being breached by smart objects. The smart devices listed above collect data on their users to ensure they are adapting effectively to their personal habits and lifestyle the best they can. This entails that data collection is performed autonomously by the smart devices and shared with each other as part of the IoT. Although this improves the effectiveness and efficiency of these devices, there may be a concern about where their personal information is being translated to outside the boundaries of their home, how it is being used, and if it is adequately protected against cyberattacks by exterior forces. The Office of the Privacy Commissioner of Canada (2020) lists some general rules of thumb to help people protect their information, encouraging homeowners to read the fine print of the devices they are putting their trust in to understand the full extent of their functions and how it affects their privacy; in many cases, certain privacy settings can be toggled on/off to block third parties (i.e., advertisers, other companies) from accessing data collected by smart home devices (Kaspersky, 2021).

Furthermore, although it might be cheaper in the long run to integrate various devices to your home, saving on costs associated with energy and water use, heating, and cooling, the upfront cost can be simply too expensive and inaccessible for some homeowners. The average cost of a fully integrated smart home can be costly to a household with a low to average income depending on how many gadgets one buys (Charlton, 2021). Traditionally, smart home tech has been advertised primarily to mid- to high-income families. Surveys also indicate that the most interest in smart devices comes from households with an income over CAN$100,000, leaving out a considerable portion of households isolated from these advancements in technology (Guidehouse Insights, 2018). A smart thermostat, for example, could cost about CAN$250 depending on the model compared with a conventional thermostat at about CAN$25 (Guidehouse, 2018). Equally, an overabundance and range of products presently available on the market is also an issue for consumers. It can become an overwhelming task to determine if your smart devices, which may be produced by different manufacturers, are compatible with each other (Caccavale, 2018). In 2018, only about 12–18% of US households were estimated to have installed a smart home device, demonstrating that there is still much work to do to get homeowners on board with the smart home and all it has to offer (Caccavale, 2018).

8.4 Residences with Home Automation

As technology continues to advance at an ever-accelerating pace, the prevalence of home automation will only increase. In a few short years, the sheer variety of home automation technology available to the everyday consumer has expanded. Moreover, the functionality of automation technology has improved to ease the daily lives of the users. The growth of automation within the home has begun to shape the genesis of home designs and planning to be more inclusive of these technologies. Below is description of a residence with integrated components of home automation.

8.4.1 BS House in Puglia, Italy

BS House was designed by Italian firm Reisarchitettura for a couple that chose a site in the small countryside of Puglia, Italy, as their home. The house functions both as a working studio and a relaxing retreat for the clients. In the design of the home, the architects drew inspiration from the traditional homes of the Puglia region, combining SHT for modern convenience.

The site sits high, with a generous view over the surrounding landscape. The house is organized around a central courtyard facing the north. As the massing of the house blocks direct sunlight, this space becomes passively cooled for enjoyable outdoor living. This spatial arrangement has been used since ancient times as protection from the harsh sun. Inside the home, the living room and workspace are to the east, while private quarters are to the west. The kitchen and dining area are in the middle, with direct access to the courtyard. A separate annex to the north provides relaxing amenities. The materiality of the home reflects traditions of Puglia. The architects used a combination of dry tone and lime plaster for the walls, with oak wood frames for openings; the design of the facades makes use of these materials in a contemporary fashion.

The clients wished to have control of their home while they are away on business trips, so a KNX automation system was integrated into the house. Not only are lighting and air-conditioning remotely adjustable, but the system also features control of the house's security system, including the entry buzzer, door lock, and alarms. Along with these elements, energy consumption management can be monitored through a smart device app. The house generates power from photovoltaic panels, which serve the home as well as a charging port for an electric car. Reisarchitettura designed the home with convenience and comfort in mind, combining traditional methods and modern technology to make the most of the site and fully embrace home automation as a tool for sustainability.

8.5 The Future of Home Automation

When automation technology first entered the consumer market, security was one of the keynote topics. Suddenly, the average consumer was able to network their entire home with their own security system. While rudimentary at first, this technology grew to include elements like motion sensors, high-pitched alarms, and eventually surveillance cameras. Home security cameras allowed users to be able to remotely monitor activity in their house to safeguard against intruders, accidents, and theft. The popularity of such technology began a trend of innovation toward creating a ubiquitous home environment.

"Smart," a moniker usually assigned to technology aimed at making daily life easier, technology has quickly grown to be a common element to many new home inventions. So-called "smart" technology infiltrates the home at a more microlevel. Smart doorbells and locks allow for users to not only record each visitor that approaches their home but allow them to monitor the status of each lock in their house from anywhere. Even lightbulbs have been reimagined to be "smart," including features like the ability to control them from one's smartphone or being able to change the light's color with the flick of a finger. The proliferation of smart home appliances aims to service people in ways that have never been seen before.

Home automation, while written off as "gimmicky," has the potential to create more accessible home environments for people with special needs, reduced mobility, or the elderly. Now, an individual who requires a wheelchair to move about their house can control their lights and thermostat through applications on their smartphone. People concerned for their senior family members can monitor their aging relatives through security systems and can make sure that their doors are properly locked without having to directly inspect them. Voice assistant technology, seen in products like the Google Home or Amazon Alexa, provides a new sense of intimacy to the world of home automation. With voice assistant software, users can search the Internet, control music or lights, send messages, and call loved ones all by simply calling out to the device. While privacy is a common concern with these technologies and they have been proven easy to hack, the benefits to this type of technology are quite significant for individuals with disabilities. This trend in bringing ubiquitous technology into the domestic sphere is marketable for companies but provides tangible help to people who truly could benefit from it.

Moreover, home automation provides new ways in which existing structures can be rewired to reduce their carbon footprint. Sensors can monitor energy usage within homes and identify ways in which reductions can be made. Furthermore, home energy systems can be aided with power-generating technology that also possesses "smart" elements. Solar panels that track the movement of the sun to ensure the highest amount of solar gain are now available to the average consumer. Sensors tracking movement can trigger lights to turn off automatically if they detect that there is no one in the room. Automation of carbon footprint-reducing activities is a great way to motivate people to purchase these innovative technologies because homeowners are able to save the environment and money on their utility bills. Other motivators include fun and engaging energy-saving games such as that described in Sect. 8.2.5. As global warming continues to impact how people live their daily lives, it is important to develop technology aimed at engraining environmental consciousness into every area where a tangible difference can be made.

Home automation will not slow down anytime soon. Architects designing new homes commonly will typically work with home automation experts to ensure that their designs can be equipped with this technology either now or in the future. Home automation technology provides users with a more intimate way to live in their own homes. Their home will directly follow the commands set forth by the individual. While many people fear home automation over privacy concerns, there are many benefits to this type of adaptive domestic technology that must be considered before being completely disregarded by architects.

Questions for a Follow-Up Discussion

1. What will you consider a smart home, and what are its key features?
2. How can digital technologies enhance a dwelling's sustainable performance?
3. What digital technologies can contribute to a home's water saving?
4. How can digital technologies contribute to a home's energy efficiency?
5. How will, in your view, a home's design be affected by future innovative digital technologies?

References

Bhardwaj, P. (2018). "Disney's 'Smart House' came out in 1999—here's all the technology from the movie that we're actually using 2 decades later." *Insider*. https://www.businessinsider.com/disney-smart-house-tech-smart-home-technologies-2018-4. Accessed April 26 2022.

Bogart, N. (2016). "Can smart thermostats really help save money on your energy bill?" *Global News*. https://globalnews.ca/news/2670317/can-smart-thermostats-really-help-save-money-on-your-energy-bill/. Accessed April 26 2022.

Bradford, A. (2021). "What are smart plugs?" *Digitaltrends*. https://www.digitaltrends.com/home/what-are-smart-plugs/

Bucher, G. (1939). The electric home of the future. *Popular Mechanical Magazine, 72*(2).

Caccavale, M. (2018). "The impact of the digital revolution on the smart home industry." *Forbes*. https://www.forbes.com/sites/forbesagencycouncil/2018/09/24/the-impact-of-the-digital-revolution-on-the-smart-home-industry/?sh=63d7a2183c76. Accessed April 26 2022.

Charlton, A. (2021). "How much does it cost to turn the average house into a smart home?" *Gearbrain*. https://www.gearbrain.com/average-smart-home-build-cost-2589554626.html. Accessed April 26 2022.

Chung, W. J., Liu, C., & Seong, Y.-B. (2019). Potential lighting and thermal demand reduction in office buildings using blind control considering surrounding buildings. *Journal of Asian Architecture and Building Engineering, 18*(3), 262–270.

Constellation. (2018). "Best ways to make a senior-friendly smart home". *Exellon*. https://blog.constellation.com/2018/07/20/smart-home-for-seniors/. Accessed April 26 2022.

de Dear, R., & Brager, G. (2021). Adaptive comfort and mixed mode conditioning. In V. Loftness (Ed.), *Sustainable built environments: Encyclopedia of sustainability science and technology series*. Springer.

Design Built Networks. (2018). "The year's biggest Design Build Network stories". https://www.designbuild-network.com/analysis/2018-the-years-biggest-design-build-network-stories/. Accessed May 18 2022.

Digiteum Team. (2021). "Key advantages of IoT (Internet of things) Water management systems." *Digiteum*. https://www.digiteum.com/smart-water-management-iot/. Accessed April 26 2022.

Fijhneer, J. D., van Oostendorp, H., Giezman, G., & Veltkamp, R. C. (2021). Competition in a household energy conservation game. *Sustainability*. https://doi.org/10.3390/su132111991. Accessed April 28 2022.

Foster, S. (2021). "10 amazing products for eco-friendly home automation." *Elemental Green*. https://elemental.green/10-amazing-products-for-eco-friendly-home-automation/. Accessed April 26 2022.

Furszyfer Del Rioa, D. D., Sovacool, B. K., & Griffiths, S. (2021). Culture, energy and climate sustainability, and smart home technologies: A mixed methods comparison of four countries. *Energy and Climate Change, 2*, 2666–2787.

Guidehouse Insights. (2018). "Smart home tech moves to low income households." https://guidehouseinsights.com/news-and-views/smart-home-tech-moves-to-low-income-households. Accessed April 26 2022.

ITA. (2021). "What is smart home irrigation?" https://integratedtechnologiesaustralia.com.au/resource-centre/what-is-smart-home-irrigation. Accessed April 26 2022.

Kaspersky. (2021). "How safe are smart homes?" *AO Kaspersky Lab*. https://www.kaspersky.com/resource-center/threats/how-safe-is-your-smart-home. Accessed April 26 2022.

Lenz, B., Schreiber, J., & Stark, T. (2011). *"Optimising building operations" in sustainable building services*. DETAIL – Institut für internationale Architektur-Dokumentation GmbH & Co. KG.

Li, J., Yang, X., & Sitzenfrei, R. (2020). Rethinking the framework of smart Water system: A review. *Water, 12*, 412.

Louis, J.-N., Caló, A., Leiviskä, K., & Pongrácz, E. (2015). Environmental impacts and benefits of smart home automation: Life cycle assessment of home energy management system. *IFAC (International Federation of Automatic Control), 48*(1), 880–885.

Office of the Privacy Commissioner of Canada. (2020). "Smart devices and your privacy." *Government of Canada*. https://www.priv.gc.ca/en/privacy-topics/technology/02_05_d_72_iot/. Accessed April 26 2022.

Sessums, Z. (2020). "We're really into the *the Jetsons'* space age style right now." *Architectural Digest: Clever*. https://www.architecturaldigest.com/story/the-jetsons-space-age-style. Accessed April 26 2022.

Sociotio. (2020). "The Facts About Ageing in Place." https://www.scioto.com/. Accessed April 26 2022.

Tuohy, J. P. (2021). "The best smart sprinkler controller". *NY Times*. https://www.nytimes.com/wirecutter/reviews/best-smart-sprinkler-controller/. Accessed April 26 2022.

U.S. Energy Information Administration. (2011). "Residential Energy Consumption Survey (RECS)." https://www.eia.gov/consumption/residential/reports/2009/air-conditioning.php. Accessed April 26 2022.

UN Water. (2021). "Water scarcity." *United Nations*. https://www.unwater.org/water-facts/scarcity/. Accessed April 26 2022.

Chapter 9
Digital Contribution to Urban Planning and Architecture

The multidisciplinary urban planning profession has evolved alongside the changing built forms and technologies. It aims to plan and regulate the use of space and its effects on economic activity and anticipate the social impact on the urban environment and its corresponding systems. Recently, the field of urban planning has embraced the use of technologies that allow for accurate visualization of urban processes and their interrelations. This chapter will define and discuss big data and its relevance to urban planning and introduce other technologies to urban planners and architects that use *urban big data* such as geographic information systems/science (GIS), three-dimensional (3D) modeling, and virtual reality (VR). The chapter will further explore these tools and how they have shaped the planning and architecture domains by allowing for a better understanding of designed spaces and how they can be improved.

9.1 Challenges That Call for Digital Means in Urban Planning

Forms of town planning have been used since the advent of cities thousands of years ago. However, the field of urban planning itself as a profession only officially emerged in the early 1900s. The field aimed to solve problems posed by urban environments that have historically and still largely concern the tasks of providing clean, comfortable, and safe environments for inhabitants while supporting economic growth, social equity, and aesthetic appeal (Fainstein, 2021). As described in Chap. 2, the concept of sustainable development introduced by a United Nations committee in the late twentieth century saw environmental sensitivity added to the agenda of urban planners. Planning processes can range in scale from master-planned cities to neighborhood design to policies. Some examples of historical urban planning movements that emerged to solve challenges posed by urbanity are as follows. In

A. Friedman, *The Sustainable Digital City*,
https://doi.org/10.1007/978-3-031-25488-8_9

1903, architect Ebenezer Howard proposed the *Garden City* concept in an attempt to manage the overcrowded and polluted industrial city (Fig. 9.1). In Howard model, each city was capped at a population size of 32,000 people with a city center for civic and entertainment activities and residences situated in clusters outside the center and surrounded by greenbelts (Moreira, 2021). In the 1950s, the suburb was introduced as a response to an influx of immigrants and soldiers returning from war and the subsequent "baby boom" that grew population sizes (Smithsonian, n.d.). The creation of the suburb and the Garden City model exemplifies "top-down" planning in which governments, private corporations, and other experts create and execute urban plans with minimal public consultation.

In her seminal book *The Death and Life of Great American Cities*, Jane Jacobs critiques many urban planning policies of the 1950s that resulted in demolition of residential sectors and the decline of the neighborhood characterized by increased segregation and violence (Jacobs, 1961). She supported "bottom-up" planning in which residents of neighborhoods have a say in decisions about place of living. Jacobs also advocated for diversity in increasingly segregated cities. She is famously known for her concept of *eyes on the street* urban planning in which neighborhoods and cities are designed to support street-level activity. This is a simple concept where having more people on city streets increases natural surveillance and in turn makes the place and them safer.

Aside from economic, social, and environmental problem-solving, urban planning is to a large degree influenced by available technologies. Transportation technologies that were outlined in Chap. 3, for example, have had an undue influence on urban morphology. The pre-industrial cities saw city shape and size accommodate foot and horse cart travel, and the streetcar in the early twentieth century resulted in a growth in city size owing to greater mobility. Next, the commuter rail era meant development along railway lines and lastly the advent of the automobile and highways filled in areas between rail lines and led to the emergence of the suburbs (Knox & McCarthy, 2005). Welwyn Garden City is an example of a planned community crossed by a rail line (Fig. 9.2). Presently, pressures of climate change and a need for sustainability have highlighted the importance of public and active transit which will no doubt affect future urban forms. The success or failure of urban planning projects can dictate the quality of a city inhabitants' life. Therefore, when proposing projects, every possible scenario must be played out to achieve the most successful outcome. Fortunately, predictive modeling is one of the many capabilities afforded to planners by GIS.

Throughout the years, the use of geographic information systems (GIS) has been key to urban planning. In its most basic form, GIS is a computer system that creates, manages, analyzes, and maps all types of geospatial data. Famously, mapping was used to identify the source of water contamination in 1854 during a cholera outbreak in London, England, at the end of the first industrial revolution, an era characterized by accelerating urbanization rates. Dr. John Snow mapped cholera outbreak locations, roads, and water sources to discover that infections originated in one of the city's water pumps. This discovery demonstrated the use of cartography as a problem-solving tool and marked the beginning of both spatial analysis and

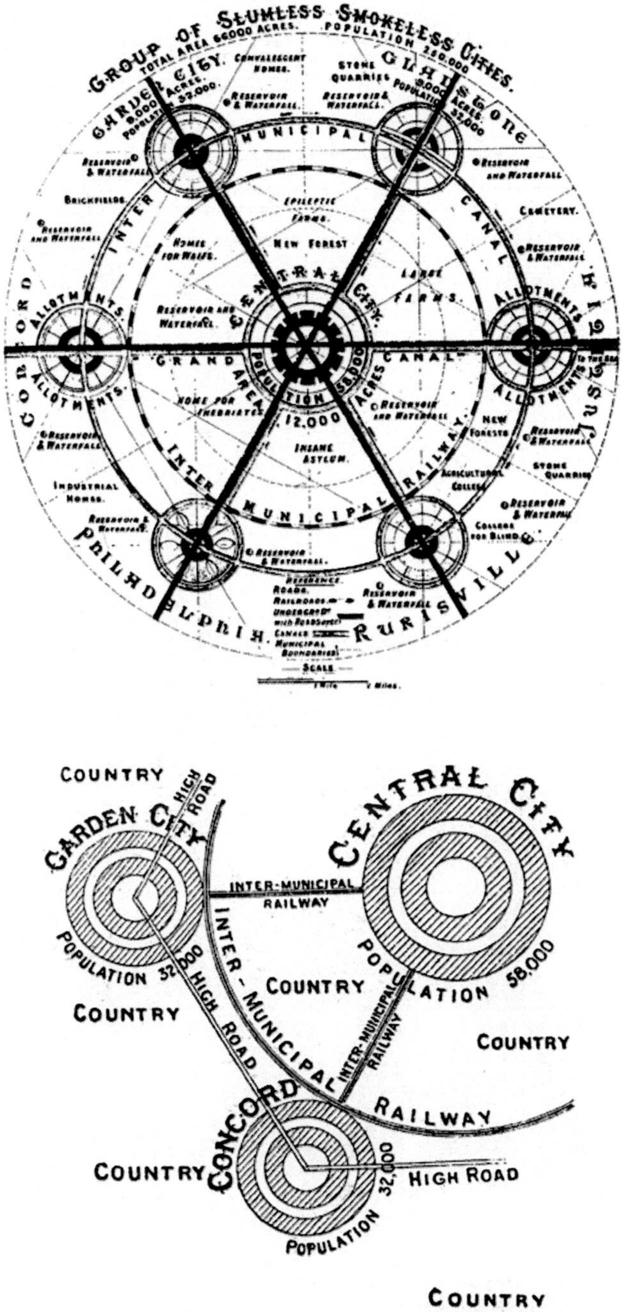

Fig. 9.1 In 1903, architect Ebenezer Howard proposed the Garden City concept to manage the overcrowded and polluted industrial city

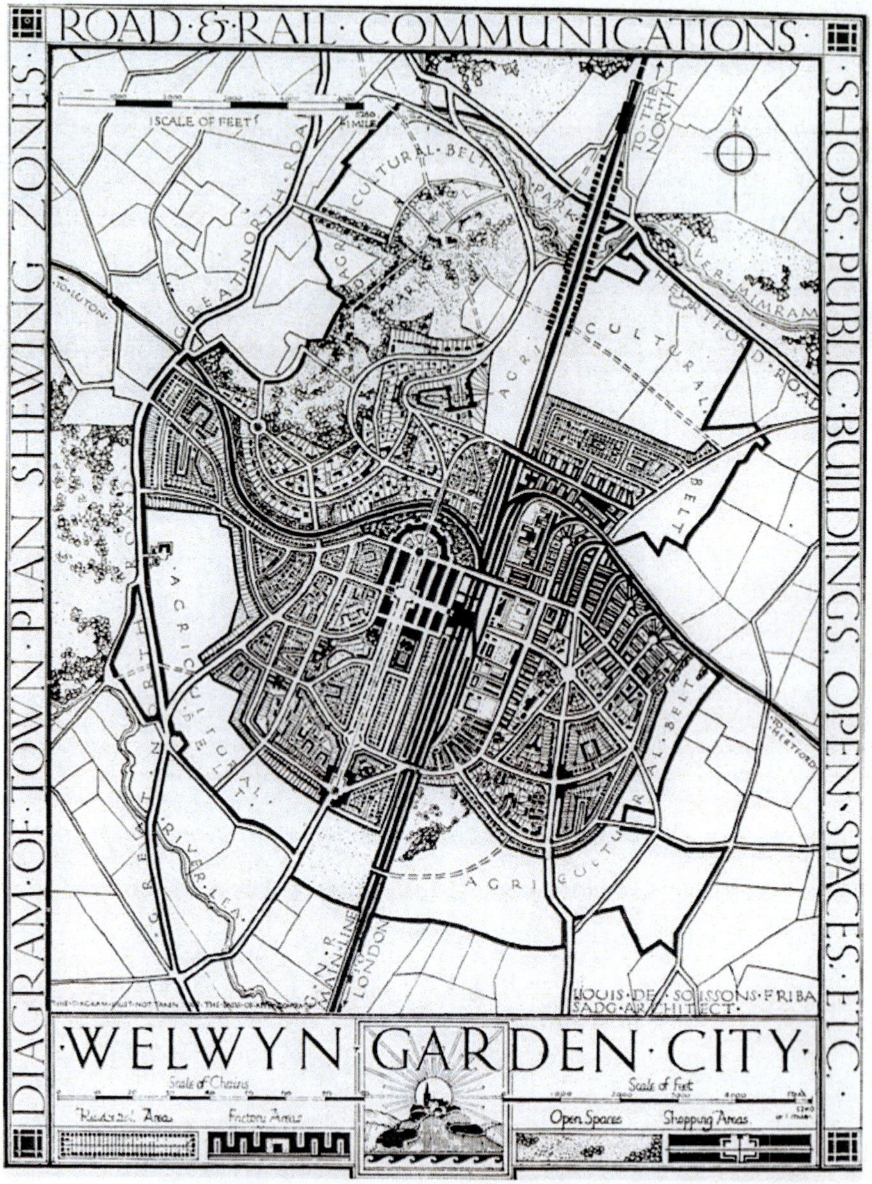

Fig. 9.2 Welwyn Garden City is an example of a planned community crossed by a rail line

epidemiology (Steenson, 2019). The next century saw little development in cartography technology as common practice for planners consisted largely of hand-drafted paper maps. Beginning in the 1960s, computer developments increased processing power and data storage, thus allowing for the documentation and calculation of coordinates. This signified a new era of modern cartographic practices characterized by the development of GIS, with visualization of census data being one of the initial uses (Fig. 9.3). Presently, as access to various data sources such as satellite imaging, crowdsourcing, and census increases, GIS technology can provide spatial data relevant to all industries. In terms of urban planning, GIS enhances visibility and gives insight into the impact and feasibility of proposed projects, as will be discussed in detail throughout this chapter.

Contemporary urban planning attempts to respond to modern challenges in ways that promote sustainable principles and pillars. Climate change calls for resilient planning that must account for rising sea levels and the increasing frequency and severity of storms. Urbanization trends mean cities need to increase density while remaining at a human scale to promote social cohesion. Aging buildings need to be updated in tasteful manners that support cultural sustainability. Constantly evolving problems require solutions of the same vein. The coming sections will explore how GIS and its aggregation and processing of increasingly complex and large set of data brings a heightened understanding of trends and solutions within cities.

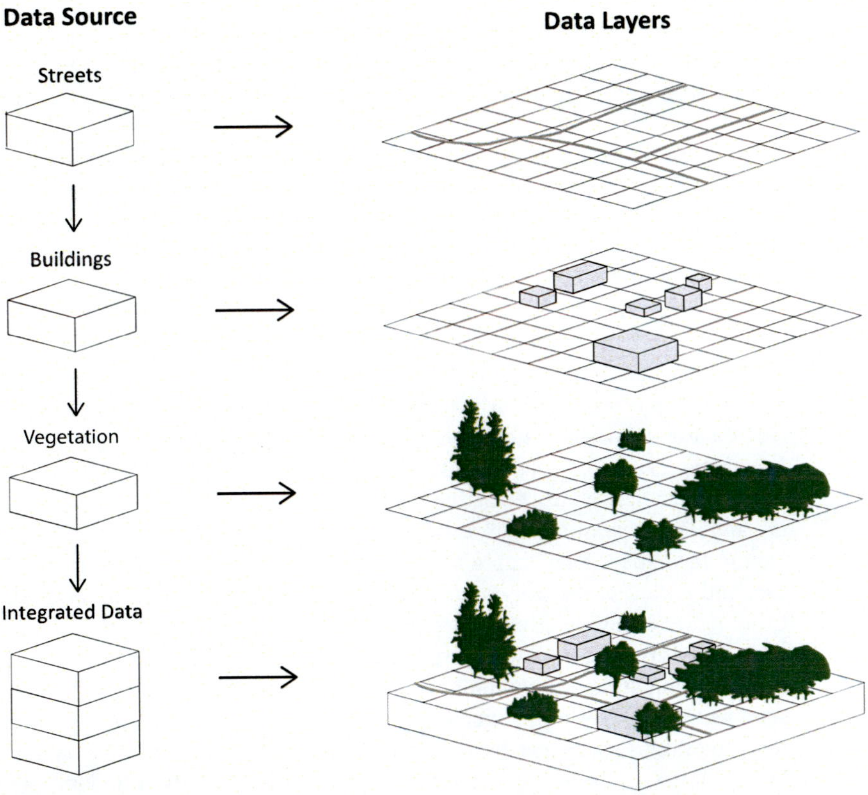

Fig. 9.3 A GIS layer model

9.2 Use of Data in Planning Cities

The growing value of data in the planning and understanding of urban systems is so prevalent that it has resulted in the rise of urban informatics. This is a new field that focuses on understanding urban systems by creating and using novel sources of *big data* (Thakuriah et al., 2017). Big data itself is more than just a buzzword and can be characterized by the 5 V's: volume, velocity, variety, veracity, and value (Douay, 2018). It describes the sum of data that is gathered through transactional, operational, planning, and social activities and interactions, it is constantly growing, and it needs to be processed and mined to be assigned value. Data mining is the extraction of valuable knowledge, trends, and correlations from large data sets to predict outcomes (SAS, n.d.). As one might imagine, predictive capabilities prove favorable to urban planners. *Metadata* on the other hand is about data, for example, metadata about a word document can include the author, title, date of last edit, file type, source, and length. Like the abstract of an academic article, metadata provides insight to the content of the data itself so that it can be efficiently sorted and identified. Metadata proves its value when information is being shared from one entity to another, but a lack of standardization can mean files of the same type contain varying forms of metadata information. While different sectors can require distinct data sets to meet their needs, the same data can be valuable to different sectors, such as census data being valuable to advertising, property management, finance, and healthcare sectors. Thakuriah et al. (2017) in *Seeing Cities Through Big Data* identify six types of "urban big data" most relevant to urban informatics and urban planners and will be discussed below.

The first and commonly associated urban big data is acquired via sensor systems. Sensors can be embedded into urban infrastructure—transportation, health, energy, water, waste, and environmental management systems, to name a few. They can also be collected remotely via sensors attached to satellites, planes, and drones. These sensors can report locations of physical infrastructures both natural and human made, the usage patterns in which people interact with infrastructure, and movement and activity patterns of populations. With the advent of the Internet of things (IoT), these sensors can now communicate machine to machine (M2M) and regulate themselves. The second type of big data outlined is *user-generated content* (UGC) characterized as social and human sensing systems. The use of crowdsourcing and the co-creation of information is one of the biggest aspects of this data collection technique. UGC can also include social media posts and comments, web use, and *global positioning system* (GPS), among others (Thakuriah et al., 2017). Community Facebook groups, for example, are an easy place to gather social, economic, and civic information on specific neighborhoods due to the nature of discussion covering current community events and happenings. Other examples are apps such as Open Street Maps where users can contribute information such as stores, road conditions, and other forms of local knowledge. Similarly, the navigation app Waze allows users to input road conditions, traffic stops, accidents, and other observed transport phenomena allowing other drivers in the area and the app to change travel routes accordingly. Gathering of these types of preexisting data is example of retroactive UCG in which researchers access past content created. Proactive UCG for researchers and planners however can come in many forms.

There are various methods, afforded by information and communications technology (ICT) in which public discussion is facilitated for feedback to public entities. Digital participation can come in many forms; the popular video game *Minecraft* in which players create a world from different voxels (3D pixels), for example, has been used to attract young people to participate in public feedback. City models are published on Minecraft, and users can edit and create the proposed infrastructure project with the ability to submit a draft to the city (Schrotter & Hürzeler, 2020). Likewise, as mentioned in Chap. 2, Singapore's policy co-creation techniques such as crowdsourcing apps and hackathons as mentioned in Chap. 2 are examples of proactive UCG. In these approaches, citizens are given the ability to directly influence and provide feedback to decision-makers, and decision-makers can ask explicit questions to gather public opinion.

The third urban big data is *administrative data* which is typically collected by public sector agencies and is derived from any type of public administrative system. It consists of healthcare, taxation, housing, vehicle licensing, education, births, deaths, national censuses, electoral registration records, and beyond (Connelly et al., 2016). The benefits of administrative data lie in its insight into information on social inequality, social policies, and human behavior as well as its ability to cover entire populations. From this, geographic determinants of health and economic/social status can be generated providing more information and value to planners. The availability and viability of administrative data sets vary nation to nation.

Private sector transaction data is the fourth urban big data and is collected by businesses through everyday transactions and used to develop customer profiles. Product purchases, terms of service agreements, and transactions through cards contribute to profiles and can be enhanced with the sensor-gathered data. Businesses use this data to improve customer services and processes; however, they can also sell this data to third parties. For example, Uber and other taxi apps can provide data showing places and times not being served through public transportations. Likewise, energy and hydro company records that are often acquired via sensors can show changing trends in consumption.

The fifth relevant urban big data source is *arts and humanities collections and historical urban data* such as archived texts, images, art, material culture, and other media. These items and their attached values can be used to guide planners on historical values and social trends to allow for culturally sustainable initiatives (Thakuriah et al., 2017). Lastly, Thakuriah et al. (2017) identify *hybrid data and linked data systems* as the final urban big data type. This is any combination of data types used to better understand desired information. Participants of social surveys, for example, may answer questions online and then wear a GPS tracker. This combines both sensor and UGC data giving researchers a broader view into the respondent's lives. While there are plenty more big data types being collected and used, narrowing down which ones are most important to planners is an important step to understanding the ubiquity of big data for the scope of this book (2017). It cannot go without mentioning that these six urban big data types are expectantly accompanied by privacy and access limitations. They are also rapidly changing the field and employment opportunities within urban planning.

Some of these changes within approaches to urban planning have been documented by Nicolas Douay (2018), author of *Urban Planning in the Digital Age*. As mentioned at the beginning of this chapter, approaches to planning range from top-down expert oriented as shown by Ebenezer Howard to bottom-up community focused later supported by Jane Jacobs that were described above. Since then, consensus has been made that a mix of the two planners, having realized the importance of both community engagement and expert opinions, is required for successful plans. Douay (2018), however, argues that with the advantages provided by big data, the "smart" urban planner is now in fact a data analyst that relies on modeling, science, and technology to define urban strategies, thereby returning to a top-down approach. While this may sound like a reversion to the early days of urban planning, the data being provided for analysis to this "smart" urban planner is embedded with public opinion in both direct and indirect ways. Big data and the unpredictability of innovative technologies have greatly changed the field of urban planning and continue to do so. The coming chapters will discuss how relevant data to urban planners is used to create, execute, and sustain urban plans.

9.3 The Use of GIS Methods and Tools

The actual use of GIS in urban planning is a rather technical process. This section will give a brief overview of some of the main components of GIS analysis by continuing the discussion on data types, giving an overview of some of the many analysis tools available, and providing some example applications. To begin, data types used in GIS are classified into either spatial or nonspatial data. Spatial data refers to data that is attributed to a specific location, such as physical objects, or weather and can be further classified as either *vector* or *raster* data (GISGeography, 2022). Vector data is made up of XY coordinates that are spatially referenced and are displayed in the forms of points, lines, or polygons. A point, for example, could represent a house where someone lives or a sensor, a line (a road), and a polygon (the square footage of a field). Raster data on the other hand consists of grids made out of pixels. Each pixel is assigned its own value or class and can represent phenomena such as elevation, land cover, or temperature. Non-inherently spatial data such as many of the previously mentioned urban big data types are often given some sort of spatial reference through a process known as *geocoding* (ArcGIS, n.d.). For example, data such as vehicle registration which is not in itself spatial is often connected to an address (either the owners or where it was registered) which can be used for spatial reference.

In terms of how spatial analyses and mapping of this data occur, there are various mapping software programs such as Maptitude, QGIS, and ArcGIS that allow users to view, manipulate, and query maps for their desired uses. These software are equipped with set tools that allow for in-depth spatial analysis. The *buffer* tool, for example, creates a buffer of a set distance around a set point. For example, if looking to build a new library but wants to ensure it is more than 2 km (1.24 miles) away from any other existing library in the city, a 2 km (1.24 mile) buffer could be created around all the libraries to ensure no overlap on further analyses. Another common tool is the "Merge" tool which allows for different data sets and their accompanying information to be merged into one layer. This could be useful, for example, when trying to analyze bars and restaurants in a city. If a planner has one file for restaurants and one file for bars, they can be merged into one so that all further analyses can be done on a single file. There is also an array of statistical analysis tools that can be used on geocoded data, often to identify spatial patterns. Tools such as these and the many others available can help urban planners determine the best location for services, recreation facilities, transport infrastructure, and zoning.

A large part of spatial analysis is making decisions about what information is relevant and what can be left out. Because models are not able to account for all aspects of the real world, GIS technicians and planners must choose the most relevant data to use when analyzing a proposed plan. For example, if looking to build a new school in a neighborhood that services the greatest number of students within a 2 km (1.24 miles) catchment area so that walking to school is feasible, a planner would need census data on who lives in that area and the available properties. A 2 km (1.24 mile) buffer would be made around all possible school locations, and these buffers then overlapped or clipped with census tract information. The planner would then be able to see which school catchment areas have the largest population of children under the age of 18 and choose the property that services the greatest number of students. There are of course many other factors that must be considered, such as accessibility and transportation, socioeconomics, and school board district boundaries, that can be added on to perform a more complex analysis to see an even bigger picture.

Another application of GIS to urban planning would be preparing for growing urban populations, for example, seeking to determine a population density level that is high enough to support local business and a healthy community but low enough to minimize environmental detriment and health risks. A planner may choose to evaluate environmental degradation levels by observing soil toxicity against population density levels. Steps of this analysis would be to first determine a population density minimum that supports a vibrant community in the determined study area. The second step would be to establish a realistic soil toxicity level cap. Third, the population density of the study area will be calculated from population size and neighborhood delineation. Finally, the planner can compare soil toxicity levels in each neighborhood first to determine if there is a relationship between soil toxicity and population density and, if there is, which levels of population density support the highest amount of community living and the lowest amounts of environmental degradation. Akin to the first example, there are factors such as location, building quality, waste management, green space, and more that can be considered to further the analyses. Examples such as these are just the beginnings of GIS applications to urban planning, as will be seen in the exploration of 3D spatial modeling in the next section.

9.4 Building and Simulating Cities Using 3D Means

Many domains that have historically relied on 2D planning are making a switch to 3D modeling. While the (X, Y) model of 2D planning makes for straightforward comprehension and has its useful instances (i.e., preliminary planning, speed, and simplicity), the three-dimensional (X, Y, Z) capabilities provided by 3D modeling means allow for increased accuracy of results and interpretations (Biljecki et al., 2015). Applications such as the estimation of solar irradiation mean buildings can be assessed for their suitability to solar panels (Fig. 9.4). Emergency response analysis means comprehensive evacuation plans, and estimation abilities such as noise and shadow prediction can be used for traffic planning and determination of solar envelopes, respectively (Biljecki et al., 2015). The ubiquitous utility of urban planning means that applications of 3D planning go far beyond those of information visualization. Yet still, within the realm of information visualization, there are a growing number of uses such as navigation visualization, flooding visualization, and visualization for communications. All this information aids planners in

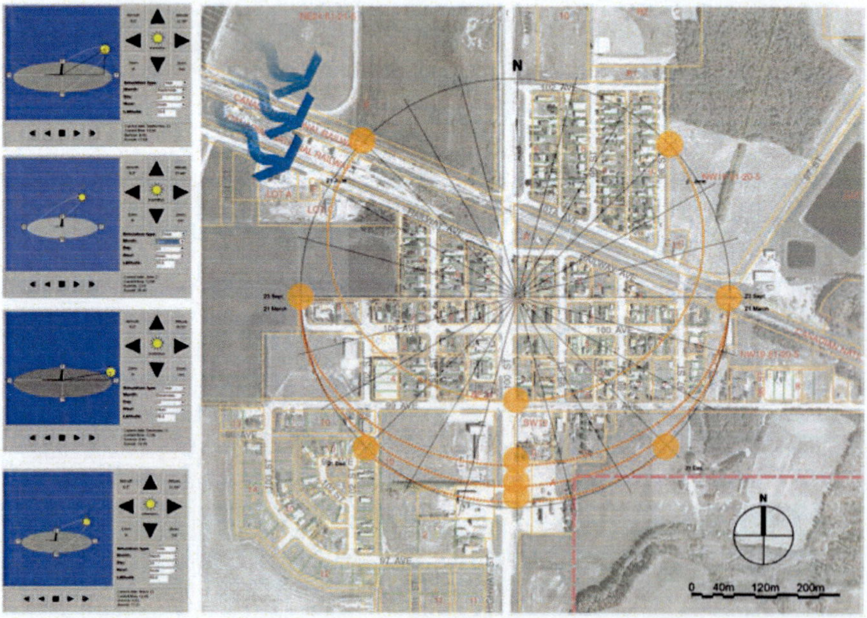

Fig. 9.4 The use of 3D modeling allows for increased planning accuracy and estimation of the solar path and buildings' suitability for the installation of photovoltaic (PV) panels

assessing situations and making decisions that maximize sustainability in all realms. Energy demand estimations and shadow predictions allow planners to situate their buildings in ways that are most conducive to natural heating or cooling depending on the conditions. They can use the most efficient materials and designs to reduce the needs for heating and cooling devices promoting environmental sustainability. Similarly, modeling the effects of predicted sea level rise on coastal cities and island nations can aid in resiliency planning. Further, the classification of building types has various applications such as marketing and real estate to ensure cultural and social sustainability. As evident, the applications of 3D modeling are far and wide. This section will expand on some of the many software facilitating these modeling capabilities.

There is a variety of 3D modeling software commonly used by urban planners that have replaced manual drafting in recent decades. These software are forms of GIS and often require raster/gridded images as inputs. AutoCAD (automated computer-aided design) is an example of a software that allows for precise 2D or 3D designs (Fig. 9.5). Similarly, SketchUp is a 3D modeling program that can turn any surface into a 3D object. It also has features such as shadow prediction at specific dates, times, and locations (Dy, 2021). Both programs come with numerous tools and extensions, 3D printing capabilities, and various supported file formats. Other software such as Rhinoceros 3D, Autodesk, and Tinker CAD can also be used for similar functions. These programs for the most part are used for individual buildings or smaller neighborhoods; there are however software such as ArcGIS CityEngine that can create "huge, interactive and immersive urban environments" using GIS and other data derived from the aforementioned software (ESRI, n.d.). CityEngine advertises building a city faster than any other program in mere five steps: choosing the area, sketching and designing, adding 3D elements, adding other details, and then exporting (ESRI, n.d.). Technologies this powerful can support meticulously detail-oriented projects such as helicopter simulations, reconstructed models of individual neighborhoods and entire cities, and creation of virtual campuses. As an example, the aging historic city of Nasiriyah in southern Iraq had been left in a state

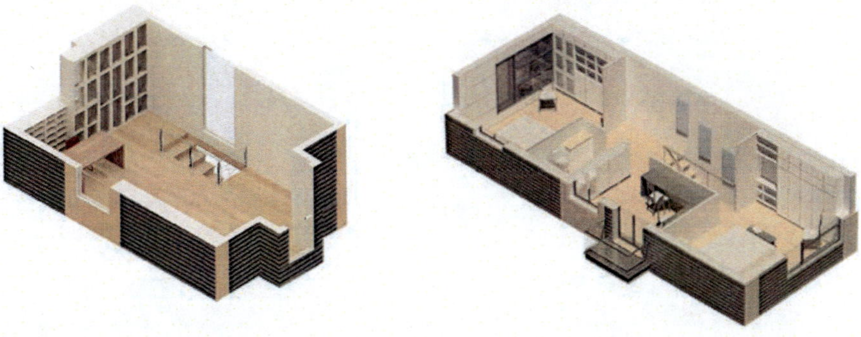

Fig. 9.5 AutoCAD (automated computer-aided design) and Rhinoceros 3D are software that allow for precise 2D or 3D designs and simulation of built environments

of repair, suffering from a lack of investment in basic city infrastructure (ESRI, 2013). A new project initiated by the Iraqi Ministry of Municipalities and Public Works aimed to create a master plan to revive and repair the city over the coming 30 years based around a culturally relevant public transit network. CityEngine allowed for the creation of a responsive model to which planners could integrate past partial models, inspiration, and future infrastructure (Fig. 9.6). Features such as vegetation models that randomly place native trees throughout the model were used to create a natural look, and various models were built to establish optimal street widths and building heights among other variables throughout the city. This concept of developing "digital twins" of cities for various uses is also popular among global cities such as Brisbane, Copenhagen, London, and Zurich (Schrotter & Hürzeler, 2020).

Another available 3D modeling technology, *building information modeling* (BIM), is an affordance of big data to architects and urban planners. While the 3D models depict much of the physical infrastructure, BIM goes one step further as to model the supporting infrastructure and function of buildings. The vast number of sensors implanted or tacked onto buildings enables the quantification and observation of water, energy, waste, noise, usage, and any other building activity to be synthesized into a single complex model (Douay, 2018). This information is valuable to all stakeholders and industries involved in a project with BIM capabilities such as architecture, construction, operation, and management. While BIM refers to individual buildings, urban planners become involved when numerous BIM equipped buildings are connected resulting in *city information modeling* (CIM) (Douay, 2018). City information models are the sum of BIM and GIS capabilities and depict detailed digital models of visible and supporting infrastructure and systems on both the neighborhood and city scale (Fig. 9.7). CIM models provide the

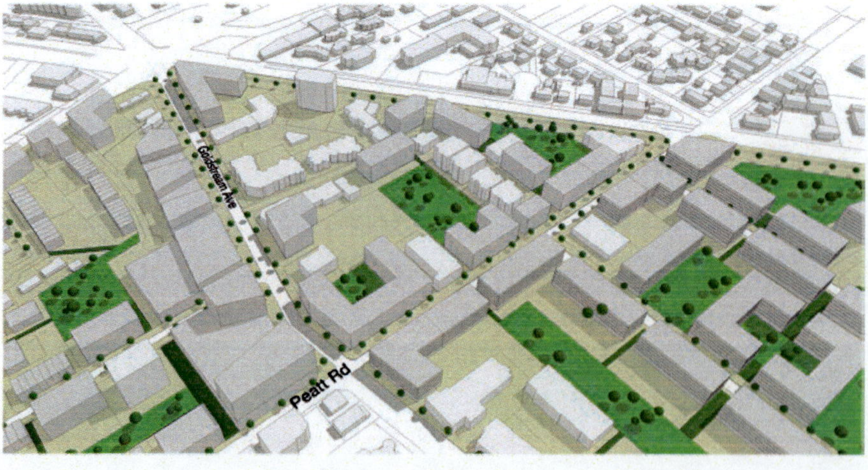

Fig. 9.6 CityEngine allowed for the creation of a responsive model to which planners could integrate past partial models, inspiration, and future structures

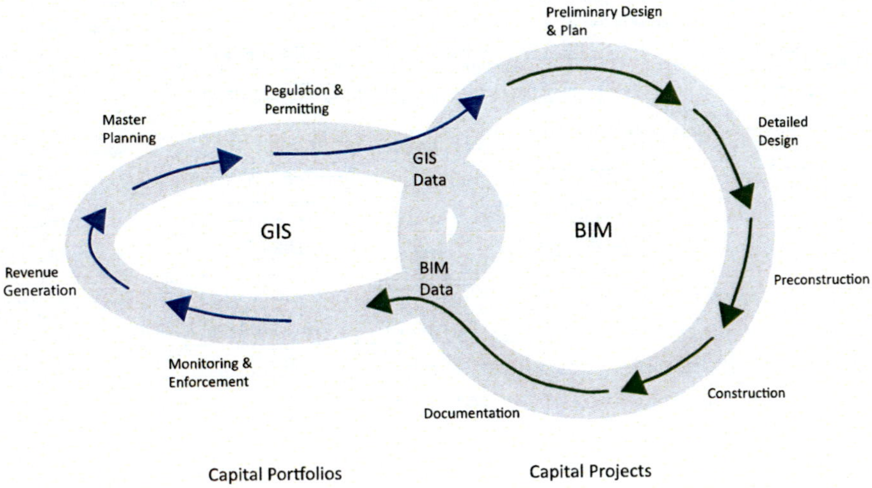

Fig. 9.7 City information models are the result of BIM and GIS capabilities and depict detailed digital models of visible and supporting infrastructure and systems on both the neighborhood and city scale

most information to support decision-making for urban planners. All buildings used in CIMs must be equipped with BIM capabilities. These models are still in their early years of development but will soon be vital tools to many urban planners among other professions.

9.5 Visualizing Future Scenarios Using Digital Means

Virtual reality (VR) is a 3D, computer-generated environment that can be explored and interacted with by a person that has been developing since the 1990s (Nguyen et al., 2016). VR is among many other rapidly growing technologies in the past 30 years and is becoming similarly ubiquitous with its ability to simulate experiences, see through "what-if" scenarios, and facilitate participatory planning (Gowling, 2019). *Augmented reality* (AR) is a variant of VR, relevant to urban planners through its capabilities to add new or different information to preexisting models. These technologies are becoming more widespread as accessibility increases and experiences can now be facilitated remotely via computer or in more immersive manners as head based like special glasses and helmets or hand-based displays (Van Dessel, 2019). By providing immersive graphics and an intuitive experience to users, VR and AR are relevant in many industries beyond urban planning such as education, healthcare, and tourism, to name a few (Nguyen et al., 2016).

VR can be used similarly to GIS and 3D modeling in many ways such as making the design process more efficient by easily comparing numerous designs, finding planning errors in early stages, and supporting multidisciplinary projects (Van

Dessel, 2019). These logistical aspects are instrumental to the design processes; VR however goes one step further by allowing for discovery of the human experience within a proposed or present design. Aspects such as walkability and noise levels throughout the day can be tested and simulated for citizens on which they can provide feedback. Further, VR can be used to conduct studies on human behavior in the urban environment to help planners design human-centric spaces (Van Dessel, 2019). VR is also a promising tool when communicating with other stakeholders such as governments and investors, who are becoming increasingly relevant actors with growing numbers of public–private partnerships.

VR can also be used for more niche applications such as seeing cities from an unfamiliar lens. For example, the *Proximity of Care Design Guide* gives planners the opportunity to experience a city at the height of a 3-year-old through VR (Proximity of Care, 2021). Through this initiative, the Design Guide encourages planners to consider children, their caregivers, and the experiential impact of the built environment. They promote the exposure of children to green spaces, accessible mobility, informal learning opportunities, and physical and social safety, among other material, policy, and social changes. As might have already been connected, designing cities to support children is beneficial to older residents as well. Using this principle, VR may also be used to see and model accessibility in cities for those who are differently abled.

9.6 What Does the Future Hold?

The importance of strategic urban planning can be seen best in its failures to execute or exist. For example, a 2021 report completed by the Polish Economic Institute discovered that Poland's inefficient urban planning has resulted in "spatial chaos" and costs the nation approximately $84 billion zloty (US$20.9 billion) annually (Wilczek, 2022). The main losses come from inefficient and congested transport, excessively dispersed housing with high costs and redundant systems that are unnecessarily maintained, and environmental costs such as built infrastructure in locations with high risks of natural disasters. The unnecessary loss of funds and incohesive planning can be felt through long commutes and difficult access to services and in the form of environmental repercussions such as poor air quality and noise pollution. The situation is not without hope; however, the report states that at least $5.8 billion zloty (US$1.4 billion) could easily be saved by optimizing transport, water, and sewage networks. There are presumably numerous political, economic, and administrative factors among others responsible for the present lack of planning. This does not however diminish Poland as an example of the consequences.

Fortunately, technology has provided undue affordances to the field and professionals of urban planning, the urban environment, its systems, and lived experiences. Lengthy and often opinion intensive planning phases that must be undergone by most urban plans can be expedited from years to months with data-driven modeling. ESRI's CityEngine boasts that an entire city can be modeled on their software in just 2 months (ESRI, n.d.). Successful urban plans have the ability to support human-centric urban environments that shape and support community networks, reduce unnecessary costs involved in planning/construction/maintenance, and embed infrastructure with environmentally conscious materials and systems. With the implementation of data-driven GIS, 3D modeling, and VR, all these capabilities can be implemented in faster and more efficient manners than ever before.

Questions for a Follow-Up Discussion

1. Name key milestones in the history of town planning and what in your view will be the next steps in the evolution of its practice?
2. What type of challenges tomorrow's cities are going to face that will call to the invention of newer digital means?
3. How did GIS contribute to the evolution of the planning profession, and how do you foresee it evolving?
4. What are the six types of "urban big data," and how will it reform decision-making in urban planning?
5. How did the use of digital tools affect the practice of architecture? What are the next steps in their evolution?

References

ArcGIS. (n.d.). "What is geocoding?". https://desktop.arcgis.com/en/arcmap/latest/manage-data/geocoding/what-is-geocoding.htm. Accessed April 26 2022.

Biljecki, F., Stoter, J., Ledoux, H., Zlatanova, S., & Çöltekin, S. (2015). Applications of 3D city models: State of the art review. *ISPRS International Journal of Geo-Information, 4*(4), 2842–2889. https://doi.org/10.3390/ijgi4042842. Accessed April 26 2022.

Connelly, R., Playford, C. J., Gayle, V., & Dibben, C. (2016). The role of administrative data in the big data revolution in social science research. *Social Science Research., 59*, 1. https://doi.org/10.1016/j.ssresearch.2016.04.015. Accessed April 26 2022.

Douay, N. (2018). "Urban planning in the digital age". Wiley. Vol. 6. Ch 1. https://onlinelibrary-wiley-com.proxy3.library.mcgill.ca/doi/book/10.1002/9781119482949. Accessed April 26 2022.

Dy, A. (2021). "The Main differences between Sketchup and AutoCAD". *Total 3D Printing*. https://total3dprinting.org/main-differences-between-sketchup-and-autocad/. Accessed April 26 2022.

ESRI. (2013). "CityEngine creates new solutions for historic cities". *ArcNews*. https://www.esri.com/about/newsroom/arcnews/cityengine-creates-new-solutions-for-historic-cities/?rmedium=arcnews&rsource=https://www.esri.com/esri-news/arcnews/spring13articles/cityengine-creates-new-solutions-for-historic-cities. Accessed April 26 2022.

ESRI. (n.d.). "ArcGIS CityEngine". https://www.esri.com/en-us/arcgis/products/arcgis-cityengine/overview. Accessed April 26 2022.

Fainstein, S. (2021). "Urban Planning". *Encyclopedia Britannica*. https://www.britannica.com/topic/urban-planning. Accessed April 26 2022.

GISGeography. (2022). "Vector vs Raster; What's the Difference Between GIS Spatial Data Types?". https://gisgeography.com/spatial-data-types-vector-raster/. Accessed April 26 2022.

Gowling, W. L. G. (2019). "How will virtual reality change the way we plan a sustainable, smart city?". https://gowlingwlg.com/en/insights-resources/articles/2019/how-will-vr-change-the-way-we-plan-smart-cities/#:~:text=VR%20serves%20as%20a%20knight,on%20infrastructure%20and%20the%20environment.&text=The%20larger%20augmented%20reality%2Fvirtual,billion%20in%20value%20by%202020. Accessed April 26 2022.

Jacobs, J. (1961). *The death and life of great American cities*. Random House.

Knox, P., & McCarthy, L. (2005). *Urbanization: An introduction to urban geography* (2nd ed.). Pearson Prentice Hall.

Moreira, S. (2021). "What are garden cities?". *Arch Daily*. https://www.archdaily.com/961275/what-are-garden-cities. Accessed April 26 2022.

Nguyen, M. T., Nguyen, H. K., Vo-Lam, K. D., Nguyen, X. G., & Tran, M. T. (2016). Applying virtual reality in City planning. In S. Lackey & R. Shumaker (Eds.), *Virtual, augmented and mixed reality* (VAMR 2016. Lecture notes in computer science, Vol. 9740). Springer. https://doi.org/10.1007/978-3-319-39907-2_69. Accessed April 26 2022.

Proximity of Care. (2021). "Design Guide". *Arup & the Bernard van Leer Foundation*. https://www.proximityofcare.com/design. Accessed April 26 2022.

SAS. (n.d.). "Data Mining History & Current Advances". https://www.sas.com/en_us/insights/analytics/data-mining.html#:~:text=Data%20mining%20is%20the%20process,relationships%2C%20reduce%20risks%20and%20more. Accessed April 26 2022.

Schrotter, G., & Hürzeler, C. (2020). The digital twin of the City of Zurich for urban planning. *PFG*, 99–112. https://doi.org/10.1007/s41064-020-00092-2. Accessed April 26 2022.

Smithsonian. (n.d.). "City and suburb". *National Museum of American History*. https://americanhistory.si.edu/america-on-the-move/city-and-suburb. Accessed April 26 2022.

Steenson, R. (2019). "The history of Geographic Information Systems (GIS)". *The Chartered Institute for IT*. https://www.bcs.org/articles-opinion-and-research/the-history-of-geographic-information-systems-gis/. Accessed April 26 2022.

Thakuriah, P., Tilahun, N. Y., & Zellner, M. (2017). Big data and urban informatics: Innovations and challenges to urban planning and knowledge discovery. In *Seeing cities through big data* (Springer Geography). Springer. https://doi.org/10.1007/978-3-319-40902-3_2. Accessed April 26 2022.

Van Dessel, A. (2019). "The use of virtual reality in urban planning". *University of Antwerp, Faculty of Design Sciences*. https://scriptiebank.be/sites/default/files/thesis/2019-09/VanDessel-Anke_Masterproef%20%28incl%20kaft%29.pdf. Accessed April 26 2022.

Wilczek, M. (2022). "Poland's "spatial chaos" caused by poor planning wastes 84bn zloty per year: Report". *Notes from Poland*". https://notesfrompoland.com/2022/01/20/polands-spatial-chaos-caused-by-poor-planning-wastes-84bn-zloty-per-year-report/?fbclid=IwAR0u6I6pU9fZGoHrgflrcl95sz81B5cxsz3sqBdqCvWGpFU5DnEQqptotDU. Accessed April 26 2022.

Chapter 10
Digital Economy and Place-Making

Billions of daily online transactions worldwide are fueling the digital economy (DE). It is a key player in the fourth industrial revolution and places high value on knowledge like big data as a good and service. The growth of the DE was recently expedited by the COVID-19 pandemic. By facilitating goods and services to be accessed from home, allowing for reduced exposure from the airborne disease, the DE proved itself invaluable. This transition from traditional in-person economies which have been the norm for centuries creates new implications for all sectors, not least urban form. This chapter will further discuss the DE, its effect on cities, the adaptation that retailers of different sizes must make, and the harms and benefits it poses to urban identity and place-making.

10.1 The Rise of the Digital Economy

Like the phenomena itself, definitions of the *DE* are constantly evolving. Presently, it is known as economic activity that results from billions of online transactions and is supported by the hyperconnectivity that evolved from the Internet, the Internet of things (IoT), and Internet communication technology (ICT) (Deloitte, n.d.). This includes money spent on infrastructure, hardware/software and their accompanying support; e-business, the digitization of services from offline to online; and goods and e-commerce, services purchased through digital means to name the original inclusions of the DE (Tattrie, 2019).

The term DE was first coined in Don Tapscott's 1995 novel *The Digital Economy: Promise and Peril in the Age of Networked Intelligence.* Tapscott predicted an economic shift away from traditional resources and toward knowledge afforded by Internet technologies, a theory which has since been proven true with the rise in the quality, quantity, and value of big data. The DE has evolved since first predicted in 1995 and is now a key factor in the fourth industrial revolution as mentioned in

A. Friedman, *The Sustainable Digital City*,
https://doi.org/10.1007/978-3-031-25488-8_10

Chap. 1. This revolution, characterized by the blurring of the lines of physical, digital, and biological realms, can be seen in some of the largest global companies today. For example, an article published by tech newspaper *TechCrunch*, in 2015, points out that Facebook, the world's most popular media owner, creates no content. In addition, Airbnb, the world's largest accommodation provider, owns no real estate (Goodwin, 2015). These companies demonstrate the power of knowledge in the DE and the interconnectivity of the physical and digital realms. It is worth keeping in mind that these data and statistics that characterize the DE are influenced largely by the global digital divide also mentioned in Chap. 1, the divide being between those that have accessible and affordable Internet and those that do not. The most recent Digital Economy Report published by the UN in 2021 found that only 20% of people in least developed countries (LDCs) use the Internet (UN, 2021). In terms of Internet usage, while eight in ten Internet users in developed countries shop online, in LDCs the same figure is less than one in ten.

Likewise, the rise of the DE, the COVID-19 contact—transmitted and airborne pandemic was also experienced differently around the world. The effects of the pandemic were more pronounced for minorities in both developed and developing countries. In another way, female owners of small businesses were 35% more likely to experience income loss compared to male counterparts—largely because women work disproportionately in industries most affected by COVID-19 (Belitski et al., 2022). Other trends include those of digital payment methods. Prior to the global pandemic which emerged in late 2019, the number of digital payments was rising, allowed by contactless payment cards, such as debit and credit cards, and online services such as payments made via mobile apps. This transition to digital payment methods is synonymous with increases in global consumption and sales (Camara, 2021).

The first few months of 2020 saw the number of online transactions grow exponentially as people were encouraged to stay home and limit the quantity and frequency of unnecessary contacts. For example, with US retailers, consumers spent $347 billion online in the first 6 months of 2020, a growth of 30% from the same period of the year prior (Bruno et al., 2020). Likewise in France, though household consumption decreased upon the emergence of the pandemic, the number of payments made by bank cards and cell phones stayed stagnant from 2019 (Camara, 2021). This was prompted by fear of contracting COVID-19 through any in-person transaction with cash being the riskiest leading to merchants refusing cash and accepting online payment methods only. These circumstances forced much of the world to purchase goods and services virtually, a shift seen most dramatically in older shoppers. This shift toward virtual banking has ultimately led to the closure of many ATMs and various physical bank branches, or in some cases branches will not reopen following closures caused by the pandemic. In Australia, for example, more than 2150 ATMs and 175 bank branches have been shut down since June 2020 (Bruno et al., 2020). As one might imagine, this trend of physical branch closure is applicable to far more industries outside of banking. Many businesses can no longer justify the expenses of maintaining in-person offices or retail sites, the consequences of which will be discussed in the coming section.

10.2 The Effects of the Shift to Digital Transactions on Physical Retail Space

Spaces for commercial transactions act both functionally for purchasing goods and services and as spaces for social gatherings, third places where people can exchange ideas and build relationships among other activities. Knowing this, one can imagine the impacts of many small- and medium-sized businesses lacking the resources to keep up with the rapid digitization of the economy, sped up by the COVID-19 pandemic. Stay-at-home orders and fear of contraction of COVID-19 have led to less foot traffic especially in downtown areas. Technology adoption has played a critical role in the survival of businesses during COVID-19, most importantly for small businesses (Abed, 2021). These are the ones however that lack the resources to develop a quality online presence or merely one at all and still must compete with big box store capabilities of delivering online services. In addition, in-store pickup style shopping, in which users shop and pay for retail items online and pick them up in stores, became a staple service that many small businesses failed to provide. A 2020 study completed by The National Bureau of Economic Research (NBER) found that of more than 5800 small businesses in the United States, 43% were expected to be closed by December 2020 (Bartik et al., 2020). Those in the hospitality, retail, personal services, entertainment, and art industries were most affected. These services are also the ones that foster much of a city's cultural sustainability. The switch to online retail, while expedited by the pandemic, has been emptying urban centers for years now. In Britain, for example, the number of empty shops rose by 7500 in 2018. Though 43,000 shops, pubs, and restaurants opened the same year, this was canceled out by the closure of 50,828 shops, resulting in a net closure rate of 37% (Clawson, 2019).

Since the pandemic, store closures have only become more frequent. In Toronto, Canada, the PATH, as described by a local newspaper *The Globe and Mail*, is a network of underground tunnels beneath the city's financial district that is lined with stores and restaurants (Moore & Younglai, 2021). The underground mall depends heavily on the thousands of Bay Street workers that, until the pandemic, passed through its facilities every weekday, making it a desirable and profitable real estate location. With major firms announcing their slow return to in-person activities following stay-at-home mandates, and/or new hybrid work schedules in which employees work part-time from home and part-time in an office, many PATH businesses cannot keep up with high rent payments. Even though the volume of people in downtown offices in Toronto has quadrupled since the pandemic low, downtown building occupancy has only reached 12% of pre-pandemic levels. University of Toronto Associate Professor Laura Miller explains that the PATH is going to have to adapt to serve needs outside of the white-collar workers upstairs, perhaps to the growing population living in downtown high rises (2021).

Despite increasing closures of independent stores, demand for goods and services is only increasing, much of which is satisfied by big box stores. With thousands of employees and entire departments dedicated to online services, retailers such as Target, Walmart, and Home Depot have all experienced huge amounts of lucrative growth since the onset of the pandemic. Target reported its greatest increase in quarterly sales in the company's history, Walmart doubled its e-commerce sales, and Home Depot increased its sales by 23% all within the first few months of the pandemic (Friedman, 2020). Similarly, the e-commerce behemoth Amazon reported, for the first 3 months of 2021, an increase in sales of 44% from 2020 (BusinessWire, 2021). While these companies are growing, it is largely in the form of e-commerce activities, which is in turn prompting the need for more warehouses and less in-person shopping space. The result is an increased number of *dark stores*, the bridge between warehouses and physical store locations. Bitterman and Hess (2021) describe the concept as resembling a typical supermarket but only open to delivery services known as *pickers* and proxy shopping services such as Postmates, Ocado, or Instacart who move quickly through the store to fill online orders. As predicted by Bitterman and Hess (2021), because dark stores are not accessible to the public, consumers will default to online shopping; similarly, dark stores do not need to be located in highly profitable and visible locations allowing for land-use changes in urban areas. As evident, both the COVID-19 pandemic and digital economies have changed consumer habits. Central markets have been a staple in society for centuries, and while this is no different today, new online economies are so powerful they are changing physical urban form. While these changes seemingly threaten third places, and long-standing sociocultural traditions, they also allow for a re-evaluation of land-use prioritization and the optimization of meeting social, economic, environmental, and cultural needs of communities.

10.3 Positive Impacts and Cases of the Changing Digital Economy

Changes of consumer habits from in-store to online shopping leave significant amounts of retail space to be reallocated. It makes up a large portion of land use throughout the world largely in affluent and suburban areas, most notably in the United States which encompasses 4.3 billion square meters (14.2 billion square feet) of retail space throughout the country. This is double that of the United Kingdom, 66% greater than Canada, and 95% greater than India (Bitterman & Hess, 2021). The shift from traditional to digital economies and the lasting impacts of COVID-19 will have many different outcomes in terms of physical space and place-making. Changing land uses and open lines of communication provided by social media and other platforms between citizens, politicians, and other stakeholders have the potential to change cities for the better. This communication fosters digital citizen involvement and participatory governance touched on throughout this book which are key to productive forward momentum in urban areas. They allow for optimization of urban branding, place-making both digitally and in-person, and human-centric urban design.

One of the major gains for urban environments with the rise of the DE is the capacity for cities to brand themselves in the form of social media campaigns. Cities are able to share their most interesting and unique features in order to attract both local and international tourists who are heavily influenced by what they see on social media and bring economic opportunity (Allam, 2020). The ability to brand a city means there must be something worth publicizing which incentivizes urban regeneration (Fig. 10.1). It increases livability and safety for residents and tourists alike and ensures a clean city by adopting pollution mitigation policies and technologies. These campaigns often begin with the consultation of citizens, tourists, stakeholders, and investors to establish what is valued in a city. Izmir, Turkey, for example, began branding in 2010 and established itself as "the frontier city," where Homer wrote the Iliad, home to Asia's first hospital, Turkey's first female theater actress, and the first Turkish city to brand itself (I Mean It Creative, 2016). Likewise, Allam (2020) describes how Amsterdam, the Netherlands, is embracing and capitalizing on its rich and artistic history. The city has installed a sign depicting the words "IAMSTERDAM" outside Rijksmuseum, the national museum of the Netherlands (Fig. 10.2). It was accompanied by the Twitter hashtag #IAMSTERDAM, and thousands of pictures with the hashtag, the sign, or both were posted on social media daily. Participatory campaigns such as these work to foster both local and global communities while also acting as advertisement for the city.

Another approach to fostering and sustaining digital urban identity and culture is the transfer of in-person activities online. Online classes, for example, are ways to bring together people with common interests and create shared experiences. Many companies offer online classes of yoga, cooking, language, computer programming, and photography. This approach of moving culture online has also been picked up by the museum industry. As described by Gheorghiu and Stefan (2018), museums act to support cultural identity by promoting material and intangible heritage and

Fig. 10.1 Key aspects of branding

Fig. 10.2 The city of Amsterdam has installed a sign depicting the words "IAMSTERDAM" outside Rijksmuseum, the national museum of the Netherlands as part of its branding effort

work to foster relationships between the visitor and exhibit materials. As of the digital age, these goals have increasingly been achieved through the use of multimedia technologies such as 3D imaging, virtual reality, and augmented reality that allow for attendees to observe artifacts in their original contexts and create more immersive experiences (2018). For example, an immersive van Gogh exhibit has been installed in cities throughout North America at which the influential artist's paintings are projected onto walls, floors, and ceilings for visitors to observe. While traditional museums are cultural staples, the inevitable trend is to offer online accommodations through the forms of virtual tours and museums themselves. *Time Maps*, for example, is a virtual museum (VM) published by the National University of Arts in Bucharest, Romania. It acts as an educational tool that highlights the identity of several lesser-known communities (Gheorghiu & Stefan, 2018). Visitors get sent a link that takes them to a map on which they can peruse different archeological sites throughout the country. Services such as these are more accessible to those with physical disability or who live far away from in-person exhibits and facilities.

As evident, there are plenty of digital opportunities for sustaining cultural identity and place-making. The COVID-19 pandemic has allowed for plenty of these technologies to be brought to society's foreground. It has also reminded many of the importance of natural everyday in-person interactions. Small talk at work, seeing the same people on daily commutes, and bumping into neighbors on the street are all interactions that foster social sustainability, even unknowingly. This craving for unmediated interactions paired with new technological abilities and shifting economies has created traction in the eyes of politicians and urban planners of the 15-min walkable neighborhood. First proposed by Carlos Moreno in 2016, it is defined as a neighborhood in which most daily needs can be met via 15-min walk or bike which allows residents to enjoy a higher quality of life. Moreno et al. (2021) further explain it as a form of chrono-urbanism which outlines that the quality of urban life is inversely proportional to the amount of time invested in transportation, more so through the use of automobiles. This concept has been furthered by other scholars to include both the 10-min walkable city and the 20-min bikeable city. Smart devices and the IoT lie at the heart of the 15-min city. Bike-sharing, car-sharing, online shopping, and drone delivery are resultant staples within these neighborhoods (Moreno et al., 2021). Various technologies providing data availability can also ensure optimal service accessibility. This data can also be used for cities and neighborhoods to brand themselves and create a place identity to increase tourism and to bolster local cultural capital. The pre-industrial architecture of many European cities such as Paris, Berlin, Milan, and Madrid allows the 15-min theory to easily be fulfilled (Creasy, 2021). Car-centric cities of North America, however, have a big task at hand reimagining the urban landscape considering present trends of excess sprawl.

Some cities are taking different approaches to chrono-urbanism. Singapore's Land Transport Authority has recently announced their goals of 20-min towns and a 45-min city (Land Transport Authority, n.d.). By walking, cycling, or public transport, citizens will be able to reach city centers within 20 min and by similar transport modes be able to get to work in peak traffic within 45 min. Likewise, Melbourne, Australia's *Plan Melbourne 2017–2050* is underpinned by the principle of a 20-min neighborhood in which daily needs can be met within a 20-min, 800-meter (0.5 mile) round trip (Victoria State Government, n.d.). Supported by research that establishes 20 min as the maximum time people are willing to walk to meet their needs, Melbourne, Australia, plans to integrate economic, transport, cultural, community, education, and recreation, among other opportunities, in every neighborhood and its city center where lanes become a tourist attraction (Fig. 10.3).

Complementary technologies to the chrono-urbanism approach also value time and aim to optimize its use. For example, the company NearSt based out of London, England, works to use digital means in support of physical stores. They provide a service with which users can search for their desired product to see if and where it is available in stores locally before they leave the house to go buy it (Clawson, 2019). NearSt connects with stores' point-of-sales (POS) systems to track in real time what it sold so users get the most accurate representation (NearSt, n.d.). They

Fig. 10.3 Melbourne, Australia, plans to integrate economic, transport, cultural, community, education, and recreation in its neighborhoods and city center where lanes are a tourist attraction

provide a service similar to in-store pickup by giving consumers the certainty that their product will be in store upon their arrival. Small businesses without the ability to provide in-store pickup services themselves can opt to partner with NearSt as a third party to improve their online presence, consumer certainty, and target local customers. The use of a program such as that offered by NearSt paired with 10-, 15-, or 20-min cities means that consumers have a brief, potentially social 10-min walk to a good or service they know for certain is available for purchase. Technology also enabled would-be home buyers to get a virtual tour of an unbuilt community and select a model as well as make a choice about the layout of a unit on its components that they are interested in and find its cost to assist them in decision-making prior to buying or renting (Fig. 10.4).

The DE has undoubtedly changed the nature of many if not all urban systems. Those that are not able to keep up with rapid technological advances are often left

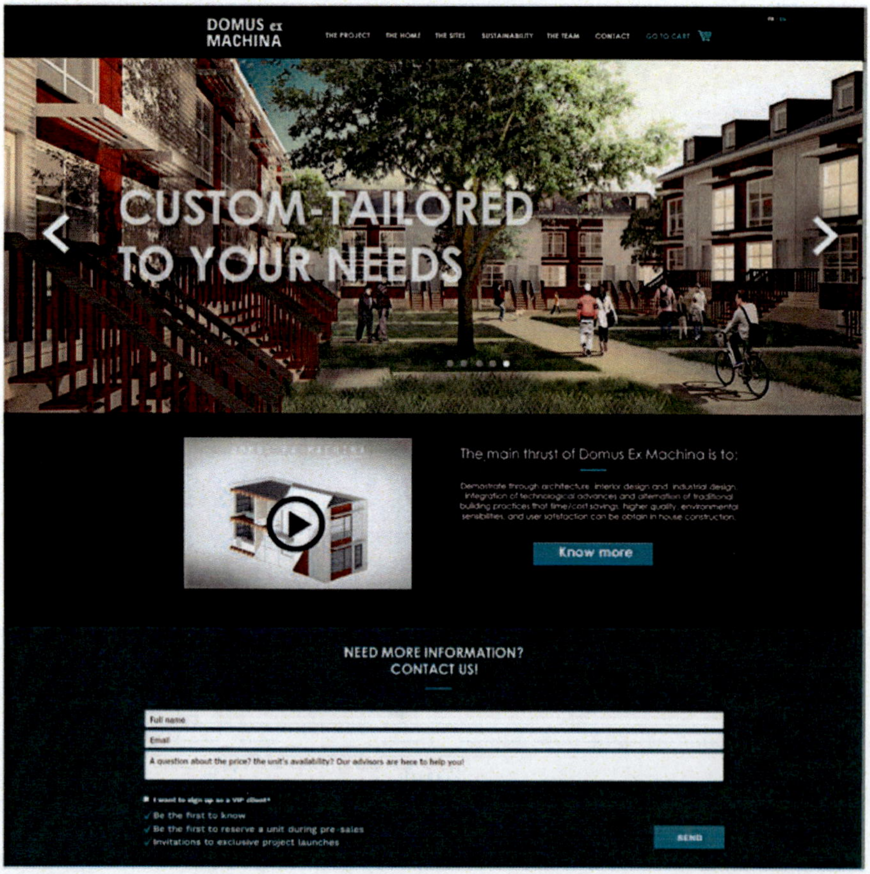

Fig. 10.4 Technologies can enable home buyers to get a virtual tour of an unbuilt community, select a model, and make a choice about the layout of a unit on its components that they are interested in and can afford

behind. The double-edged sword of these technologies is displayed in this chapter. Online shopping means that practically all goods and services are available at the tap of a keyboard but has also threatened many "mom and pop" retailers that are community staples. Social media can allow for cities to brand themselves in a way that they choose to be seen yet can also be used maliciously to discredit competition. It is for these reasons that the communication abilities provided by emerging technologies are some of the most important. They can be used to keep everybody informed and up to date on technological and urban changes and allow for as many voices as possible to be heard.

Questions for a Follow-Up Discussion

1. What are the positive and negative effects of the DE on a city's retail activities?
2. How can digital means assist in the urban renewal of a rundown city district?
3. How can digital means contribute to the branding and promotion of cities?
4. In what ways digital technologies can strengthen and expand local cultural activities and use of cultural hubs?
5. How can digital technologies support the concept of 15-min cities?

References

Abed, S. S. (2021). A literature review exploring the role of technology in business survival during the covid-19 lockdowns. *International Journal of Organizational Analysis, 20210518*. https://doi.org/10.1108/IJOA-11-2020-2501. Accessed April 26 2022.

Allam, Z. (2020). Digital urban networks and social media. In *Cities and the digital revolution*. Palgrave Pivot. https://doi.org/10.1007/978-3-030-29800-5_3. Accessed April 26 2022.

Bartik, A., Bertrand, M., Cullen, Z. B., Glaeser, E. L., Luca, M., & Stanton, C. T. (2020). How are small businesses adjusting to COVID-19? Early evidence from a survey. *National Bureau of Economic Research*. https://www.nber.org/papers/w26989. Accessed April 26 2022.

Belitski, M., Guenther, C., Kritikos, A. S., & Thurik, R. (2022). Economic effects of the COVID-19 pandemic on entrepreneurship and small businesses. *Small Business Economics, 58*, 593–609. https://doi.org/10.1007/s11187-021-00544-y. Accessed April 26 2022.

Bitterman, A., & Hess, D. B. (2021). Going dark: The post-pandemic transformation of the metropolitan retail landscape. *Town Planning Review, 92*, 385–394. https://doi.org/10.3828/tpr.2020.57. Accessed April 26 2022.

Bruno, P., Denecker, O., & Niederkon, M. (2020). "The accelerating winds of change in global payments". *The 2020 McKinsey Global Payments Report*. McKinsey & Company. https://www.mckinsey.com/~/media/mckinsey/industries/financial%20services/our%20insights/accelerating%20winds%20of%20change%20in%20global%20payments/2020-mckinsey-global-payments-report-vf.pdf. Accessed April 26 2022.

Business Wire. (2021). "Amazon announces first quarter results". https://www.businesswire.com/news/home/20210429006037/en/Amazon.com-Announces-First-Quarter-Results. Accessed April 26 2022.

Camara, Y. (2021). "Digital payments and business resilience evidence in the time of COVID-19". *World Bank Group*. https://openknowledge.worldbank.org/bitstream/handle/10986/35613/Digital-Payments-and-Business-Resilience-Evidence-in-the-Time-of-COVID-19.pdf?sequence=1&isAllowed=y. Accessed April 26 2022.

Clawson, T. (2019). "Online shopping is killing physical stores—Can a digital platform come to the rescue?". Forbes. https://www.forbes.com/sites/trevorclawson/2019/07/28/online-shopping-is-killing-physical-stores-can-a-digital-platform-come-to-the-rescue/. Accessed April 26 2022.

Creasy, A. (2021). "Perspective on the 20-minute neighbourhood—progress around the world". *LGIU Local Government Information Unit*. https://lgiu.org/perspectives-on-the-20-minute-neighbourhood-progress-around-the-world/#:~:text=Over%20the%20last%20ten%20years,or%20local%20public%20transport%20trip%E2%80%9D. Accessed April 26 2022.

Deloitte. (n.d.). "What is digital economy?". https://www2.deloitte.com/mt/en/pages/technology/articles/mt-what-is-digital-economy.html. Accessed April 26 2022.

Friedman, G. (2020). "Big-box retailers' profits surge as pandemic marches on". *The New York Times*. https://www.nytimes.com/2020/08/19/business/coronavirus-walmart-target-home-depot.html. Accessed April 26 2022.

Gheorghiu, D., & Stefan, L. (2018). "Virtual museums: Dealing with cultural identity in the digital age". https://www.researchgate.net/publication/324151100_VIRTUAL_MUSEUMS_DEALING_WITH_CULTURAL_IDENTITY_IN_THE_DIGITAL_AGE. Accessed April 26 2022.

Goodwin, T. (2015). "The Battle is for the customer Interface". *Tech Crunch*. https://techcrunch.com/2015/03/03/in-the-age-of-disintermediation-the-battle-is-all-for-the-customer-interface/. Accessed April 26 2022.

I Mean It Creative. (2016). "Izmir City branding—Case study". https://www.youtube.com/watch?v=yCWTh8mQK3w&ab_channel=IMeanItCreative. Accessed April 26 2022.

Land Transport Authority. (n.d.). "Land transport master plan 2040". https://www.lta.gov.sg/content/ltagov/en/who_we_are/our_work/land_transport_master_plan_2040.html. Accessed April 26 2022.

Moore, O., & Younglai, R. (2021). "Toronto's PATH system faces crisis as companies announce permanent hybrid work arrangements". *The Globe and Mail*. https://www.theglobeandmail.com/business/article-torontos-path-system-faces-crisis-as-companies-announce-permanent/. Accessed April 26 2022.

Moreno, C., Allam, Z., Chabaud, D., Gall, C., & Pratlong, F. (2021). Introducing the "15-Minute City": Sustainability, resilience and place identity in future post-pandemic cities. *Smart Cities, 4*(1), 93–111. https://doi.org/10.3390/smartcities4010006. Accessed April 26 2022.

NearSt. (n.d.). https://near.st/. Accessed April 26 2022.

Tattrie, J. (2019). "Canada and the digital economy". The Canadian Encyclopedia. https://www.thecanadianencyclopedia.ca/en/article/digital-economy-in-canada. Accessed April 26 2022.

UN. (2021). "Digital Economy Report 2021". United Nations Conference on Trade and Development. https://unctad.org/webflyer/digital-economy-report-2021. Accessed 10 April 2022.

Victoria State Government. (n.d.). "20-minute neighbourhoods". https://www.planning.vic.gov.au/policy-and-strategy/planning-for-melbourne/plan-melbourne/20-minute-neighbourhoods. Accessed April 26 2022.

Chapter 11
Future Digital Cities

Digital technology is and can further be woven into urban fabrics to make cities more sustainable given present and future social challenges. Impacts of this can be seen from large-scale urban systems such as energy management and virtual healthcare to small everyday changes such as installing smart home appliances and riding a shared bike. This chapter will reflect on the many examples of the ways in which cities can be made more comfortably livable, efficient, and sustainable in all its meanings through the adoption of various technologies.

11.1 Reflections

This book has outlined ways in which digital technology is and can further be woven into urban fabrics to make cities more sustainable given present and future social challenges. Growing population and urbanization rates shine light on the importance of sound supply chains, efficient transportation methods, and affordable housing. The worsening effects of climate change such as changing temperatures, rising sea levels, and increasing frequency and severity of extreme weather events demonstrate time and again the importance of resilient infrastructure and long-term planning. Lastly, phenomena such as pandemics highlight and exacerbate the digital divide as well as inequalities in access to healthcare, insurance, and job security, among other support systems. These events create challenges but also provide unique opportunities. The COVID-19 pandemic, for example, acted as a catalyst for many industries to further adopt technologies they otherwise might have gone years without using. Healthcare facilities have more broadly adopted telehealth capabilities. Originally to support physical distancing, telehealth also allows for easier and faster access to healthcare professionals and reduces unnecessary travel. Similarly, the videoconferencing platform, which also supported physical distancing measures, allows for people to meet face-to-face from anywhere with an IoT device.

A. Friedman, *The Sustainable Digital City*,
https://doi.org/10.1007/978-3-031-25488-8_11

This means less travel is required for business, lowering air traffic, vehicle traffic, and GHG emissions and bettering air quality. More on well-being and sustainable transportation in digital cities can be found in Chaps. 4 and 7, respectively.

This adoption of technology—which had been an occurrence long before the COVID-19 pandemic—also allows for furthering of sustainable practices and policies. The pillars of sustainability, social, cultural, economic, environmental, and governmental, can all be uniquely supported by digital means, and likewise can the principles of sustainability be supported; the path of least negative impact, self-sustaining processes, supporting relations, and the life cycle approach. GIS technology, for example, allows for city expansion to be centered around the human experience, addressing all the pillars of sustainability by quantifying the accessibility of proposed projects and taking as many built and environmental factors into account as possible. Relatedly, digital marketplaces work to support numerous sustainable pillars and practices. These platforms give used items a second life prolonging their life span and reducing ecological impact, sustain local barter and trade economies, foster relations, and create communities among buyers and sellers. Likewise, promotional emails and social media provide platforms for cultural information and social cohesion. Restaurants and shops have the ability to brand themselves for free as they see fit and reach a wide audience, and likewise to other services, festivals, and events that promote local culture and values and strengthen communities. More on the foundations of sustainability can be explored in Chap. 2.

This book expands on the examples above and gives a broad scope of technological applications that are increasingly present in our lives and can bolster sustainability within the city. These exist within urban systems as big as energy distribution, urban planning, and the digital economy and as small as home automation applications, digital healthcare, and bike-riding. The guidelines on which this book is set are largely influenced by the 11th Sustainable Development Goal (SDG) proposed by the UN in 2015 which underlines the need for inclusive, safe, resilient, and sustainable cities and communities. Accompanying targets of this goal are to be met by 2030 and include those such as universal access to safe, inclusive, and accessible green and public spaces, safe access to affordable and sustainable transport systems, improving road safety, and reducing the environmental impact of cities among other environmental, economic, social, cultural, and governance goals (UN, n.d.). We also aim to have highlighted the power of the human role in actively adopting these technologies to be used for good and in the creation of problems that technology is used to solve.

11.2 Sharing Knowledge

The present tech boom has been characterized as the fourth industrial revolution. It blurs the lines between the physical and digital realms and places a high value on knowledge as a good. Growing availability of Internet communication technologies (ICTs) means that knowledge sharing is easier than ever before. ICT paired with globalization means that foreign knowledge and solutions to urban problems are more accessible and can be shared on a global scale. For centuries, urbanists have been modeling cities in hopes of understanding them and sharing knowledge. Consider the concentric zone model, originally theorized by sociologist Ernest Burgess, attempts to characterize the growth of a city. Burgess describes a central business district as the heart of a city, surrounded by a light industrial zone, a low-income housing zone, and so forth (Burdett, 2018). While no one city or model can act as a cookie-cutter template to all others, ideas and inspiration for new ways to integrate sustainability into cities can now be shared around the world with the click of a button. As technology has allowed for our needs to be met differently, the physical design of cities reflects that. Technology allows cities to implement different forms of urban plans, governance, and social services and communicate what works and what does not based on a multitude of factors. Various data collection abilities also allow for municipalities to gather real-time data on the effectiveness of implemented changes.

There are numerous online resources for sharing solutions to urban challenges. For example, the World Economic Forum (WEF) Global Agenda Council on the Future of Cities compiled a list of ten urban innovations from around the world. This includes resources such as home- and vehicle-sharing platforms, intelligent streetlamps equipped with sensors, and water pipes connected to the IoT to optimize and identify issues in water pipe networks (World Economic Forum, 2015). Online library catalogs and research centers also serve as platforms for academic and other research to be shared. A study published by Sim et al. (2015) identified that the success of a city can in part be measured by the number of face-to-face, opportunity spawning interactions between its inhabitants. Aside from the social and cultural benefits of increasing connectivity within a city, it can also boost GDP as proven by Sim et al. (2015). Likewise, Milton et al. (2021) published an article in the *Journal of Transport and Health* which used a natural experiment to evaluate the effectiveness of 30 km/h (20mp/h) speed limits. They used the cities of Edinburgh and Belfast, two cities which had already implemented 30 km/h (20 mp/h) limits on many non-arterial roads and were able to establish that these lower speed limits reduce collisions and casualties from their higher counterparts. This reduces stress on healthcare systems and makes them an effective public health intervention. Traffic speed is particularly important for pedestrians and bikers in case of accidents; therefore, lowering it makes for safer biking and walking in cities. More on sustainable transportation and interventions allowed by technology can be found in Chap. 3. Other cities such as Melbourne experiment with solutions to urban problems within themselves. The Resilient Melbourne Strategy (RMS), with the help of more than 1000 individuals, identified services and places in Melbourne that lack resiliency and aim to deliver, organize, and coordinate urban services and infrastructures across a variety of contexts such as mobility, energy, housing, health, climate change, and social cohesion in new and more robust ways (Fastenrath & Coenen, 2021). This strategy has also been dubbed one of experimental governance which incorporates participatory programs and addresses different stakeholders. Published articles such as these can act as baselines for planners and policy makers and are widely available with a quick Internet search. They provide plenty of direction and motivation to implement systems of participatory governance that create more face-to-face interactions and reduce city speed limits which in turn can increase safety, economic activity, and community ties, all with minimal intervention.

This knowledge-sharing ability can also help inform planners on what not to do moving forward. For example, the Sidewalk Labs smart city project proposed for the city of Toronto, Canada's waterfront was to be a pioneer in the realm of the affordable and sustainable built environment. It was going to be built entirely out of timber, have a high bikeability and wide sidewalks, and feature both affordable housing and business space (Gibson, 2019). The main issue surrounding this project arose around the topic of data governance. Being a smart city, data was to be collected from many different sources within the project such as home thermostats and road sensors. As a result, an "urban data trust" was proposed, a body of public and private actors who would manage data collected from the project's sensors (Austin & Lie, 2021). The uncharted waters of data collection within public–private partnerships (PPP) highlighted the outdated laws in Canada regarding data access and privacy which were therefore unable to address the issues. The Sidewalk Labs Toronto project came to a halt in the spring of 2020 near the onset of the COVID-19 pandemic. However, issues around data security and privacy were never resolved and constantly evoked rumors of lack of transparency between parties. Sidewalk Labs Toronto serves as a lesson to planners and civil leaders alike on the importance of communication and updated policies when it comes to data privacy and PPPs.

All the above examples align with the goals of the 11th SDG and its targets. Research, cities themselves, and proposed projects such as the ones mentioned here as well as other tools such as conferences all serve as communicable experiments on how best to urban plan moving forward. Because there are no one-size-fits-all cities, these experiments can be changed based on a city's history, geographic positioning, economic activity, and population, among other factors. These experiments are happening globally and, thanks to ICT, results on how the pillars of sustainability are affected as well as public opinion can easily be shared and evaluated for fit elsewhere.

11.3 Creating and Integrating Systems

Assuming the rapid intake of new technology persists into the future, there is no doubt that many urban systems will integrate their online platforms. While today urban services and places are digitizing at varying rates, some industries such as the media and ICT have seemingly gone fully digital, whereas others such as construction and entertainment have more non-digital components; the future will most likely see more digitization. Aside from solving problems and generating wealth, technology aims to make life more convenient, and one platform that can deliver many services is the most convenient of them all. Many public transit apps, for example, not only cover bus and subway times and accessibility but have expanded to incorporate car-share and bike-share services into their systems. Integrated modeling is also a way of the future. This book has covered different modeling platforms that are working to integrate as many systems as possible into one. The intersection of geographic information systems (GIS) and building information modeling (BIM) will allow for many environmental and human effects of proposed projects to be visualized and quantified in a single model.

Existing cities adopt technology at their own paces. They do the same with the digitization and integration of services. Many new master-planned cities (NMPCs), on the other hand, which are being built to accommodate rapid urbanization, are built from scratch with nearly all systems digitized and integrated. NEOM in Saudi Arabia as described by Bhawsar and Madakam (2021) is an NMPC slated to finish its US$500 billion construction by 2025. The city will be the first of its kind that will be fully automated and digital. This will be facilitated by drones, smartphones, self-driving cars, robots, and AI, among other IoT devices. The city will have all the services we have come to expect in a city—offices, healthcare facilities, stadiums, entertainment parks, and more—but will have robots completing all manual work and be entirely powered by renewable energy sources (2021). The benefits and shortcomings of high-tech NMPCs are plenty. Planners can design a city that fits perfectly into its environmental surroundings; they do not have to worry about outdated systems getting in the way of upgrades and can accommodate many more people. Downfalls include inflexible plans making bumps along the road difficult to handle and the lack of a shared and cultural history. Whether it be NMPC or old ones adopting new technologies, the streamlining of services is in the future. This might be through conglomerate companies or simply stakeholders working together; a few platforms offering more services will be the most efficient and convenient.

11.4 Safety of Digital Systems

As the amount of IoT devices continues to increase, so do their surrounding cybersecurity issues. One of the few certainties going forward into the digital age is that cybercrime will continue to innovate and grow (Deloitte, n.d.), although simultaneously growing are the abilities and applications of artificial intelligence (AI) such as machine learning (ML) and deep learning (DL) to defend against malware and ransomware among other cyberattacks. Education on cybercrime is also an important tool that can be implemented to protect against other cyber risks such as gaming and gambling addictions, cyber addictions, and personal information exposure (Rahman et al., 2020). Educating the youth on these risks is key to creating responsible Internet users as their use of the Internet is constantly changing and increasing. The ability to protect people, their devices, and their data from cyberattacks will increase public trust in data collection agents and allow for more data to be safely collected resulting in more efficient and tailored services.

Alongside the technological and educational opportunities to defend against cybersecurity attacks are changing and updating regulations. Many governments were late to the game in terms of cyber protections; the Government of Canada, for example, only developed a national effort to defend against cybersecurity threats in 2010. In instances such as these, technological developments are numerous steps ahead of legislation. As cyberattacks increase in frequency and severity along with the amount of IoT devices, governments and businesses will have no choice but to update cyber defense abilities and policies. The same lack of legislation goes for protection of personal privacy and data. For years, companies have been selling and monetizing user data under ambiguous circumstances with uncertain consent. The European Union (EU) has been a pioneer in terms of data security with member nations such as Norway, Denmark, and Sweden having some of the most secure Internet privacy regulations in the globe (Grant, n.d.). Most recently, in 2018, the EU updated the General Data Protection Regulation (GDPR) which is a legal framework that requires businesses—such as banks and insurance companies—to protect personal data and privacy collected from an EU citizen within an EU state (Rossow, 2018). Any company within the EU or any that interacts with its citizens that processes or stores personal data must comply with the GDPR. There are 99 articles outlined in the GDPR that companies must comply with such as the right for user data to be erased as requested, clear opt-in/opt-out notices, creating and enacting a data breach incident response plan, and hiring a data protection officer (Rossow, 2018). As the EU has been a leader in terms of data protection, it can be predicted that countries around the world will follow suit in terms of data protection regulations. Legislation such as this will provide clarity regarding privacy to data users and providers. Required transparency will hopefully make it so that data security issues such as those that arose around the Sidewalk Labs Toronto project are matters of the past.

11.5 Final Thoughts

One of the many messages put forward by this book is that the human aspect of technology cannot be forgotten. Humans will always have a role to play whether it be setting up ML algorithms or fixing hardware. Now that this plethora of information is available at our fingertips, we must ensure that it is safely and securely processed, analyzed, and shared. We must also work to close gaps such as that presented by the digital divide. IoT devices are becoming more affordable and accessible making this plausible and beneficial to everyone allowing for data collected to better represent global population needs. Moreover, as different sectors and countries adopt new technologies, the powers and systems that we know and are comfortable with today will change. Cryptocurrency, for example, is changing the present banking system with which we are familiar. It is a much larger change than any in recent history such as the introduction of debit cards, credit cards, and ATMs. Change is practically the only constant in the sustainable digital city, and we must strive to ensure people and cities are empowered by the introduction of new technologies.

Questions for a Follow-Up Discussion

1. Which social domains are likely to see rapid expansion in the use of digital technologies?
2. Should the rapidly expanding role of digital technologies be a concern to people (i.e., losing jobs)?
3. Does the collection of data about citizens pose a privacy concern? In what way?
4. Can digital technologies contribute to achieving the UN Sustainable Development Goals (SDGs)? How?
5. What possible integration between different digital technologies and applications do you foresee taking place?

References

Austin, L., & Lie, D. (2021). Data trusts and the governance of smart environments: Lessons from the failure of sidewalk labs urban data trust. *Surveillance & Society, 19*, 255. https://ojs.library.queensu.ca/index.php/surveillance-and-society/article/view/14409/9783 Accessed April 29 2022.

Bhawsar, P., & Madakam, S. (2021). NEOM Smart City: The City of future (The Urban Oasis in Saudi Desert). In J. C. Augusto (Ed.), *Handbook of smart cities*. Springer. https://doi.org/10.1007/978-3-030-15145-4_86-2. Accessed April 26 2022.

Burdett, M. (2018). "Urban land use patterns and models". *GeographyCaseStudy*. https://www.geographycasestudy.com/urban-land-use-patterns-and-models/. Accessed April 26 2022.

Deloitte. (n.d.). "Future of cyber". https://www2.deloitte.com/global/en/pages/about-deloitte/articles/gx-future-of-cyber.html. Accessed April 26 2022.

Fastenrath, S., & Coenen, L. (2021). Future-proof cities through governance experiments? Insights from the Resilient Melbourne Strategy (RMS). *Regional Studies, 55*(1), 138–149. https://doi.org/10.1080/00343404.2020.1744551. Accessed April 26 2022.

Gibson, E. (2019). "Sidewalk labs reveals masterplan for sidewalk Toronto Neighbourhood". *Dezeen*. https://www.dezeen.com/2019/06/28/toronto-tomorrow-sidewalk-labs-toronto/#. Accessed April 29 2022.

Grant, A. (n.d.). "Internet privacy by country". *Best VP*. https://bestvpn.org/privacy-index/. Accessed April 26 2022.

Milton, K., Kelly, M. P., Baker, G., Cleland, C., Cope, A., Craig, N., Foster, C., Hunter, R., Kee, F., Kelly, P., Nightingale, G., Turner, K., Williams, A. J., Woodcock, J., & Jepson, R. (2021). Use of natural experimental studies to evaluate 20mph speed limits in two major UK cities. *Journal of Transport & Health, 22*. https://doi.org/10.1016/j.jth.2021.101141. Accessed April 26 2022.

Rahman, N. A. A., Sairi, I. H., Zizi, N. A. M., & Khalid, F. (2020). The Importance of Cybersecurity Education in School. *International Journal of Information and Education Technology, 10*. https://doi.org/10.18178/ijiet.2020.10.5.1393. Accessed April 26 2022.

Rossow, A. (2018). "The birth of GDPR: What is it and what you need to know". *Forbes*. https://www.forbes.com/sites/andrewrossow/2018/05/25/the-birth-of-gdpr-what-is-it-and-what-you-need-to-know/?sh=5de41a4955e5. Accessed April 26 2022.

Sim, A., Yaliraki, S. N., Barahona, M., & Stumpf, H. (2015). Great cities look small. *Journal of the Royal Society, Interface, 12*, 20150315. https://doi.org/10.1098/rsif.2015.0315. Accessed April 26 2022.

UN. (n.d.). "Goal 11: Make cities inclusive, safe, resilient and sustainable". *United Nations*. https://www.un.org/sustainabledevelopment/cities/. Accessed April 26 2022.

World Economic Forum. (2015). "Top ten urban innovations". https://www.weforum.org/agenda/2017/02/cities-must-tirelessly-innovative-to-respond-to-their-challenges/. Accessed April 26 2022.

Illustration Credits

Figures not listed below are in the public domain or have been conceived, drawn or photographed by the author and/or members of his research and design teams. Their names are listed in the Acknowledgements and the Projects' Teams pages. Every effort has been made to list all contributors and sources. In case of omission, the author and the publisher will include appropriate acknowledgement or correction in any subsequent edition of this book.

Fig. 1.4 After Parks Associates. Trends in Smart Home Security and Privacy.
Fig. 2.1 After OECD (2012), OECD Environmental Outlook to 2050.
Fig. 4.2 After Statistics Canada Labour Force Survey and Labour Force Survey supplement, February 2021.
Fig. 4.3 After Statistics Canada Labour Force Survey and Labour Force Survey supplement, February 2021.
Fig. 9.3 After National Geographic 2022.
Fig. 9.6 After Shimonti 2018.

A. Friedman, *The Sustainable Digital City*,
https://doi.org/10.1007/978-3-031-25488-8

Bibliography

Abdulla, M. A., & Esmaeel, A. M. (2019). Providing information through smart platforms: An applied study on academic libraries in Saudi universities. *Journal of Education, Society and Behavioural Science*, 1–24.

Abed, S. S. (2021). A literature review exploring the role of technology in business survival during the covid-19 lockdowns. *International Journal of Organizational Analysis, 20210518*, 1045. https://doi.org/10.1108/IJOA-11-2020-2501. Accessed April 26 2022.

Abnett, K., & Jessop, S. (2021). "U.S., Canada among 20 countries to commit to stop financing fossil fuels abroad". *Reuters*. https://www.reuters.com/business/cop/19-countries-plan-cop26-deal-end-financing-fossil-fuels-abroad-sources-2021-11-03/. Accessed April 14 2022.

Aeris. (2017). "Connected IoT Aviation and Airplane Monitoring". https://cdn2.hubspot.net/hubfs/246745/Aeris%20Industry%20Brief%20-%20Smart%20Airplane%20Monitoring.pdf. Accessed 6 April 2022.

Ahmed, Z., Ali, S., & Wasif, Z. M. (2020). Linking urbanization, human capital and the ecological footprint in G7 countries: An empirical analysis. *Sustainable Cities and Society, 55*, 102064. https://doi.org/10.1016/j.scs.2020.102064. Accessed 10 April 2022.

Ahvenniemi, H., et al. (2017). What are the differences between sustainable and smart cities? *Cities, 60*(Part A). https://doi.org/10.1016/j.cities.2016.09.009. Accessed 6 April 2022.

Alalouch, C., Bouzguenda, I., & Fava, N. (2019). Toward smart sustainable cities: A review on the role of digital citizen participation could play in advancing social sustainability. *Sustainable Cities and Society, 50*, 101627. https://doi.org/10.1016/j.scs.2019.101627. Accessed 10 April 2022.

Al-Arshani, S. (2020). "School districts across the country are using school buses to deliver wifi to students who lack access". Business Insider. https://www.insider.com/wifi-buses-being-used-across-country-to-give-kids-internet-2020-3. Accessed April 26 2022.

AlertReady.ca. (n.d.). Alert ready: Emergency alert system. https://www.alertready.ca/. Accessed April 13 2022.

Ali, T., Irfan, M., Alwadie, A. S., & Glowacz, A. (2020). IoT-based smart waster bin monitoring and municipal solid waste management system for smart cities. *Arabian Journal for Science and Engineering, 45*. https://doi.org/10.1007/s13369-020-04637-w. Accessed April 14 2022.

Allam, Z. (2020). Digital urban Networks and social media. In *Cities and the digital revolution*. Palgrave Pivot. https://doi.org/10.1007/978-3-030-29800-5_3. Accessed April 26 2022.

American Psychological Association. (n.d.). "How do i choose between medication and therapy?" from https://www.apa.org/ptsd-guideline/patients-and-families/medication-or-

A. Friedman, *The Sustainable Digital City*,
https://doi.org/10.1007/978-3-031-25488-8

therapy#:~:text=Research%20generally%20shows%20that%20psychotherapy,improve%20outcomes%20from%20psychotherapy%20alone. Accessed April 14 2022.

American Public Transportation Association. (2016). "Shared mobility and the transformation of public transit: Research analysis." https://www.apta.com/wp-content/uploads/Resources/resources/reportsandpublications/Documents/APTA-Shared-Mobility.pdf. Accessed April 10 2022.

Amin, S. R., Tamina, U., & Jimenez, L. A. (2017). Understanding air pollution from induced traffic during and after the construction of a new highway: Case study of highway 25 in montreal. *Journal of Advanced Transportation, 2017*, 1–14. https://doi.org/10.1155/2017/5161308. Accessed April 10 2022.

ArcGIS. (n.d.). "What is geocoding?" https://desktop.arcgis.com/en/arcmap/latest/manage-data/geocoding/what-is-geocoding.htm. Accessed April 26 2022.

Arnett, H. (2017). The challenges in quantifying social sustainability at a neighborhood level. *Cities & Health, 1*(2), 139–140. https://doi.org/10.1080/23748834.2018.1460104. Accessed 10 April 2022.

Asch, D. A., Patel, M. S., & Volpp, K. G. (2015). Wearable devices as facilitators, not drivers, of health behavior change. *JAMA, 313*(5), 459–460.

Aunan, K., Chowdry, S., Lelieveld, J., & Venter, Z. S. (2020). COVID-19 lockdowns cause goobal air pollution declines. *Earth, Atmosphere, and Planetary Sciences., 117*, 18984. https://doi.org/10.1073/pnas.2006853117. Accessed April 14 2022.

Austin, L., & Lie, D. (2021). Data trusts and the governance of smart environments: Lessons from the failure of sidewalk labs urban data trust. *Surveillance & Society*. https://ojs.library.queensu.ca/index.php/surveillance-and-society/article/view/14409/9783 Accessed April 29 2022, *19*, 255.

Bajya, P., Choudhary, M., Kaswan, S., Kaswan, V., & Kumar, P. (2019). Green production strategies. In *Encyclopedia of food security and sustainability*. Elsevier. https://doi.org/10.1016/B978-0-08-100596-5.22292-0. Accessed 10 April 2022.

Bankmycell. (n.d.). "How many smartphones are in the world". https://www.bankmycell.com/blog/how-many-phones-are-in-the-world. Accessed 6 April 2022.

Bartik, A., Bertrand, M., Cullen, Z. B., Glaeser, E. L., Luca, M., & Stanton, C. T. (2020). "How are small businesses adjusting to COVID-19? Early evidence from a survey". National Bureau of Economic Research. https://www.nber.org/papers/w26989. Accessed April 26 2022.

Belitski, M., Guenther, C., Kritikos, A. S., & Thurik, R. (2022). Economic effects of the COVID-19 pandemic on entrepreneurship and small businesses. *Small Business Economics, 58*, 593–609. https://doi.org/10.1007/s11187-021-00544-y. Accessed April 26 2022.

BetterHelp. (n.d.). "About". https://www.betterhelp.com/about/. Accessed April 14 2022.

Bhardwaj, P. (2018). "Disney's 'smart house' came out in 1999 – here's all the technology from the movie that we're actually using 2 decades later." Insider. https://www.businessinsider.com/disney-smart-house-tech-smart-home-technologies-2018-4. Accessed April 26 2022.

Bhawsar, P., & Madakam, S. (2021). NEOM Smart City: The City of future (The urban oasis in Saudi Desert). In J. C. Augusto (Ed.), *Handbook of smart cities*. Springer. https://doi.org/10.1007/978-3-030-15145-4_86-2. Accessed April 26 2022.

Biljecki, F., Stoter, J., Ledoux, H., Zlatanova, S., & Çöltekin, S. (2015). Applications of 3D city models: State of the art review. *ISPRS International Journal of Geo-Information, 4*(4), 2842–2889. https://doi.org/10.3390/ijgi4042842. Accessed April 26 2022.

Bitterman, A., & Hess, D. B. (2021). Going dark: The post-pandemic transformation of the metropolitan retail landscape. *Town Planning Review, 92*, 385–394. https://doi.org/10.3828/tpr.2020.57. Accessed April 26 2022.

Bixi. (n.d.). "Pricing". https://www.bixi.com/en/pricing. Accessed April 14 2022.

Bloom, N., Liang, J., Roberts, J., & Ying, Z. J. (2015). Does working from home work? Evidence from a Chinese experiment. *The Quarterly Journal of Economics, 130*(1), 165–218. https://www.gsb.stanford.edu/faculty-research/publications/does-working-home-work-evidence-chinese-experiment Accessed April 13 2022.

Bogart, N. (2016). "Can smart thermostats really help save money on your energy bill?" Global News. https://globalnews.ca/news/2670317/can-smart-thermostats-really-help-save-money-on-your-energy-bill/. Accessed April 26 2022.

Bollen, M. H. (2017a). "The power grid as part of a 100% renewable energy system". *Large Scale Grid Integration of Renewable Energy Sources*. Chap. 1, pp. 1–27. IET Digital Library. https://doi-org.proxy3.library.mcgill.ca/10.1049/PBPO098E_ch1. Accessed April 14 2022.

Bollen, M. H. (2017b). "The power grid as part of a 100% renewable energy system". *Large Scale Grid Integration of Renewable Energy Sources*. Chap. 1, pp. 1–27. IET Digital Library. https://doi-org.proxy3.library.mcgill.ca/10.1049/PBPO098E_ch1. Accessed 6 April 2022.

Bouton, S., Knupfer, S. M., Mihov, I., & Swartz, S. (2015). "Urban mobility at a tipping point." McKinsey Sustainability. https://www.mckinsey.com/business-functions/sustainability/our-insights/urban-mobility-at-a-tipping-point. Accessed April 10 2022.

Bradford, A. (2021). "What are smart plugs?" Digitaltrends. https://www.digitaltrends.com/home/what-are-smart-plugs/. Accessed April 26 2022.

Bravo, E. (2019). "Amsterdam launches a modular urban lighting system". Tomorrow. City. https://tomorrow.city/a/amsterdam-launches-a-modular-urban-lighting-system. Accessed April 26 2022.

Brody, J. (2017). "Social interaction is critical for mental and physical health". https://www.nytimes.com/2017/06/12/well/live/having-friends-is-good-for-you.html. Accessed April 26 2022.

Bruno, P., Denecker, O., & Niederkon, M. (2020). "The accelerating winds of change in global payments". The 2020 McKinsey Global Payments Report. McKinsey & Company. https://www.mckinsey.com/~/media/mckinsey/industries/financial%20services/our%20insights/accelerating%20winds%20of%20change%20in%20global%20payments/2020-mckinsey-global-payments-report-vf.pdf. Accessed April 26 2022.

Bucher, G. (1939). The Electric home of the future. *Popular Mechanical Magazine, 72*(2).

Budget Direct. (n.d.). "Car Ownership Singapore 2022". *Budget Direct Insurance*. https://www.budgetdirect.com.sg/car-insurance/research/car-ownership-singapore#:~:text=Highlights,million%20vehicles%20on%20Singapore's%20roads. Accessed 10 April 2022.

Burdett, M. (2018). "Urban land use patterns and models". GeographyCaseStudy. https://www.geographycasestudy.com/urban-land-use-patterns-and-models/. Accessed April 26 2022.

BusinessWire. (2021). "Amazon announces first quarter results". https://www.businesswire.com/news/home/20210429006037/en/Amazon.com-Announces-First-Quarter-Results. Accessed April 26 2022.

Caccavale, M. (2018). "The impact of the digital revolution on the smart home industry." Forbes. https://www.forbes.com/sites/forbesagencycouncil/2018/09/24/the-impact-of-the-digital-revolution-on-the-smart-home-industry/?sh=63d7a2183c76. Accessed April 26 2022.

Calder, K. E. (2016). "Singapore: Smart city, smart state". Brookings Institution Press. Chap. 5. https://muse-jhu-edu.proxy3.library.mcgill.ca/chapter/1883643. Accessed 6 April 2022.

Camara, Y. (2021). "Digital payments and business resilience evidence in the time of COVID-19". World Bank Group. https://openknowledge.worldbank.org/bitstream/handle/10986/35613/Digital-Payments-and-Business-Resilience-Evidence-in-the-Time-of-COVID-19.pdf?sequence=1&isAllowed=y. Accessed April 26 2022.

Campbell, A., & Tindall, A. (2010). "Water Security: National and Global Issues". *U.S. Geological Survey*. https://pubs.usgs.gov/fs/2010/3106/pdf/FS10-3106.pdf. Accessed 10 April 2022.

Canada. (2017). "About renewable Energy". https://www.nrcan.gc.ca/our-natural-resources/energy-sources-distribution/renewable-energy/about-renewable-energy/7295. Accessed April 14 2022.

Canada. (2020). "Net-zero: Future building standards". https://www.nrcan.gc.ca/energy-efficiency/energy-efficiency-homes/buying-energy-efficient-new-home/netzero-future-building-standards/20581. Accessed April 14 2022.

Canada. (2022). "Net-zero emissions by 2050". https://www.canada.ca/en/services/environment/weather/climatechange/climate-plan/net-zero-emissions-2050.html. Accessed April 14 2022.

Canada Energy Regulator. (2020). "Canada's renewable power landscape 2017 – Energy market analysis". Canada Energy Regulator. https://www.cer-rec.gc.ca/en/data-analysis/energy-commodities/electricity/report/2017-canadian-renewable-power/canadas-renewable-power-landscape-2017-energy-market-analysis-ghg-emission.html. Accessed 6 April 2022.

Canadian Commission for UNESCO. (2020). "Connecting culture and sustainability. https://en.ccunesco.ca/blog/2020/1/connecting-culture-and-sustainability. Accessed 10 April 2022.

Canadian Medical Association. (2019). "Virtual care in Canada: Discussion paper". *CMA Health Summit*. https://www.cma.ca/sites/default/files/pdf/News/Virtual_Care_discussionpaper_v2EN.pdf. Accessed April 15 2022.

Caradonna, L. (2018). "Part III historiography of sustainability". *Routledge international handbooks.* Ch in Routledge handbook of the history of sustainability. https://books.google.ca/books?hl=en&lr=&id=CoI6DwAAQBAJ&oi=fnd&pg=PT14&dq=history+of+sustainability&ots=1LG7U9JxcL&sig=XyrBQclQT8gOES2-SEjGv8w3tus#v=onepage&q=history%20of%20sustainability&f=false. Accessed 10 April 2022.

Center for Climate and Energy Solutions. (n.d.). "Global Emissions". https://www.c2es.org/content/international-emissions/#:~:text=Most%20of%20the%20world's%20greenhouse,the%20United%20States%20and%20Russia. Accessed April 14 2022.

Center for Disease Control (CDC). (2020). "Loneliness and social isolation linked to serious health conditions". https://www.cdc.gov/aging/publications/features/lonely-older-adults.html. Accessed April 15 2022.

Charlton, A. (2021). "How much does it cost to turn the average house into a smart home?" Gearbrain. https://www.gearbrain.com/average-smart-home-build-cost-2589554626.html. Accessed April 26 2022.

Chen, G.-M., & Sawyer, R. (2012). "The Impact of Social Media on Intercultural Adaptation". University of Rhode Isand. https://digitalcommons.uri.edu/com_facpubs/15/. Accessed April 28 2022.

Chen, Z., van Lierop, D., & Ettema, D. (2020). Dockless bike-sharing systems: What are the implications? *Transport Reviews, 40*(3), 333–353.

Chung, W. J., Liu, C., & Seong, Y.-B. (2019). Potential lighting and thermal demand reduction in office buildings using blind control considering surrounding buildings. *Journal of Asian Architecture and Building Engineering, 18*(3), 262–270.

City Bike. (n.d.). "Pricing". https://www.citibikenyc.com/pricing. Accessed April 14 2022.

Clawson, T. (2019). "Online shopping is killing physical stores – Can a digital platform come to the rescue?" Forbes. https://www.forbes.com/sites/trevorclawson/2019/07/28/online-shopping-is-killing-physical-stores-can-a-digital-platform-come-to-the-rescue/. Accessed April 26 2022.

Clean Energy Grid. (n.d.). "Our Outdated Grid". *Americans for a Clean Energy Grid.* https://cleanenergygrid.org/our-outdated-grid/. Accessed April 14 2022.

Cloud, R. (2014). "A virtual retirement community". https://retirementandgoodliving.com/a--virtual-retirement-community/. Accessed April 14 2022.

College of Physiotherapists of Ontario. (n.d.). "Virtual Practice". https://www.collegept.org/registrants/virtual-practice-in-physiotherapy. Accessed April 14 2022.

Commauto Montreal. (n.d.). "How It works". https://montreal.communauto.com/?lang=en&city=montreal. Accessed April 25 2022.

Connelly, R., Playford, C. J., Gayle, V., & Dibben, C. (2016). The role of administrative data in the big data revolution in social science research. *Social Science Research, 59*. https://doi.org/10.1016/j.ssresearch.2016.04.015. Accessed April 26 2022.

Constellation. (2018). "Best ways to make a senior-friendly smart home". Exellon. https://blog.constellation.com/2018/07/20/smart-home-for-seniors/. Accessed April 26 2022.

Contento, A. (2018). "Water scarcity: The Most understated global security risk". *Harvard Law School National Security Journal.* https://harvardnsj.org/2018/05/water-scarcity-the-most-understated-global-security-risk/#_edn11. Accessed 10 April 2022.

Cook, J. E., & Doyle, C. (2002). Working alliance in online therapy as compared to face-to-face therapy: Preliminary results. *Cyberpsychology & Behavior, 5*(2), 95–105.

Creasy, A. (2021). "Perspective on the 20-minute neighbourhood – progress around the world". LGIU Local Government Information Unit. https://lgiu.org/perspectives-on-the-20-minute-neighbourhood-progress-around-the-world/#:~:text=Over%20the%20last%20ten%20years,or%20local%20public%20transport%20trip%E2%80%9D. Accessed April 26 2022.

Creative Energy. (n.d.). "District Energy". http://creative.energy/district-energy. Accessed April 14 2022.

Crosbie, T., & Moore, J. (2004). Work-life balance and working from home. *Social Policy and Society, 3*(3), 223.

D'Orso, G., Migliore, M., & Caminiti, D. (2020). The environmental benefits of carsharing: The case study of Palermo. *Transportation Research Procedia, 48*, 2127–2139. https://doi.org/10.1016/j.trpro.2020.08.271. Accessed April 10 2022.

Davies, A. (2016). "Copenhagen's new traffic lights recognize and favor cyclists." Wired. https://www.wired.com/2016/02/copenhagens-new-traffic-lights-recognize-and-favor-cyclists/. Accessed April 10 2022.

de Dear, R., & Brager, G. (2021). Adaptive comfort and mixed mode conditioning. In V. Loftness (Ed.), *Sustainable built environments: Encyclopedia of sustainability science and technology series*. Springer.

DEC. (n.d.). "What is DE?" Decentralized Energy Canada DEA. https://www.deassociation.ca/whatisde. Accessed April 14 2022.

Deloitte. (n.d.-a). "Future of cyber". https://www2.deloitte.com/global/en/pages/about-deloitte/articles/gx-future-of-cyber.html. Accessed April 26 2022.

Deloitte. (n.d.-b). "What is digital economy?" https://www2.deloitte.com/mt/en/pages/technology/articles/mt-what-is-digital-economy.html. Accessed April 26 2022.

Dennison, L., Morrison, L., Conway, G., & Yardley, L. (2013). Opportunities and challenges for smartphone applications in supporting health behavior change: Qualitative study. *Journal of Medical Internet Research, 15*(4), e86.

Design Built Networks. (2018). "The year's biggest Design build network stories". https://www.designbuild-network.com/analysis/2018-the-years-biggest-design-build-network-stories/. Accessed May 18 2022.

Devlin, J. F., Yap, N. T., & Weir, R. (2005). Public participation in environmental assessment: Case studies on EA legislation and practice. *Canadian Journal of Development Studies/Revue canadienne d'études du développement, 26*(3), 487–500.

Diao, M., Kong, H., & Zhao, J. (2021). Impacts of transportation network companies on urban mobility. *Nature Sustainability, 4*, 494–500. https://doi.org/10.1038/s41893-020-00678-z. Accessed April 10 2022.

Dietsch, D. (2008). "Live/work: Working at home, living at work". Abrams.

Digiteum Team. (2021). "Key advantages of IoT (internet of things) Water management systems." Digiteum. https://www.digiteum.com/smart-water-management-iot/. Accessed April 26 2022.

Douay, N. (2018). "Urban planning in the digital age". Wiley. Vol. 6. Ch 1. https://onlinelibrary-wiley-com.proxy3.library.mcgill.ca/doi/book/10.1002/9781119482949. Accessed April 26 2022.

Doxy.me. (n.d.). "For patients". https://doxy.me/en/patients/. Accessed April 14 2022.

Dutzik, T., Madsen, T., & Baxandall, P. (2013). "A new way to go: The transportation apps and vehicle-sharing tools that are giving Americans the freedom to drive less". U.S. PIRG Education Fund, Frontier Group. https://uspirg.org/sites/pirg/files/reports/A%20New%20Way%20to%20Go%20vUS1_1.pdf. Accessed April 13 2022.

Dwivedi, Y., Hughes, L., Ismagilova, E., & Rana, N. P. (2020). "Security, Privacy and Risks Within Smart Cities: Literature Review and Development of a Smart City Interaction Framework". Information Systems Frontiers. https://link.springer.com/article/10.1007/s10796-020-10044-1. Accessed 6 April 2022.

Dy, A. (2021). "The Main differences between Sketchup and AutoCAD". Total 3D Printing. https://total3dprinting.org/main-differences-between-sketchup-and-autocad/. Accessed April 26 2022.

Ecolizer. (n.d.). https://www.ecolizer.be/. Accessed April 14 2022.

Enbridge Gas. (n.d.). "Renewable Hydrogen". https://www.enbridgegas.com/sustainability/renewable-hydrogen. Accessed April 14 2022.

Energy.gov. (n.d.). "Solar integration: Inverters and grid services basics". *Office of Energy Efficiency & Renewable Energy.* https://www.energy.gov/eere/solar/solar-integration-inverters-and-grid-services-basics#:~:text=An%20inverter%20is%20one%20of,which%20the%20electrical%20grid%20uses. Accessed April 14 2022.

Environmental Protection Department [EPD]. (2015). "3D tools facilitate public participation in EIA process". http://www.epd.gov.hk/eia/englisremoh/press_release/3D.htm

EPA. (2017). "Geothermal Energy". Environmental Protection Agency EPA. Archive. https://archive.epa.gov/climatechange/kids/solutions/technologies/geothermal.html

EPA. (n.d.). "Sources of greenhouse Gas emissions." EPA, Environmental Protection Agency. https://www.epa.gov/ghgemissions/sources-greenhouse-gas-emissions. Accessed 6 April 2022.

ESRI. (2013). "CityEngine creates new solutions for historic cities". ArcNews https://www.esri.com/about/newsroom/arcnews/cityengine-creates-new-solutions-for-historic-cities/?rmedium=arcnews&rsource=https://www.esri.com/esri-news/arcnews/spring13articles/cityengine-creates-new-solutions-for-historic-cities. Accessed April 26 2022.

ESRI. (n.d.). "ArcGIS CityEngine". https://www.esri.com/en-us/arcgis/products/arcgis-cityengine/overview. Accessed April 26 2022.

Evans-Cowley, J., & Hollander, J. (2010). The new generation of public participation: Internet-based participation tools. *Planning Practice and Research, 25*, 397–408. https://doi.org/10.1080/02697459.2010.503432

Fainstein, S. (2021). "Urban planning". Encyclopedia Britannica. https://www.britannica.com/topic/urban-planning. Accessed April 26 2022.

FAO. (2021). "The State of Food Security and Nutrition in the World 2021". Food and Agriculture Organization of the United Nations. https://www.fao.org/state-of-food-security-nutrition. Accessed 10 April 2022.

Fastenrath, S., & Coenen, L. (2021). Future-proof cities through governance experiments? Insights from the Resilient Melbourne Strategy (RMS). *Regional Studies, 55*(1), 138–149. https://doi.org/10.1080/00343404.2020.1744551. Accessed April 26 2022.

Fijhneer, J. D., van Oostendorp, H., Giezman, G., & Veltkamp, R. C. (2021). Competition in a household Energy conservation game. *Sustainability, 13*. https://doi.org/10.3390/su132111991. Accessed April 28 2022.

Fishman, T. D., & Flynn, M. (2018). "Part two: Funding and financing smart cities series." Using Public-Private Partnerships to Advance Smart Cities; Deloitte Center for Government Insights: London, UK, 1–10. https://www2.deloitte.com/content/dam/Deloitte/global/Documents/Public-Sector/gx-ps-public-private-partnerships-smart-cities-funding-finance.pdf. Accessed 6 April 2022.

Fleming, S. (2019). "Traffic congestion cost the US economy nearly $87 billion in 2018." World Economic Forum. https://www.weforum.org/agenda/2019/03/traffic-congestion-cost-the-us-economy-nearly-87-billion-in-2018/. Accessed April 10 2022.

Foster, S. (2021). "10 amazing products for eco-friendly home Automatiom." Elemental Green. https://elemental.green/10-amazing-products-for-eco-friendly-home-automation/. Accessed April 26 2022.

Frazer, J. (2019). "How the gig economy is reshaping careers for the next generation". Forbes. https://www.forbes.com/sites/johnfrazer1/2019/02/15/how-the-gig-economy-is-reshaping-careers-for-the-next-generation/?sh=7738282549ad. Accessed April 13 2022.

Freedman, M. (2021). "How businesses are collecting data (and what They're doing with It)". *Business News Daily.* https://www.businessnewsdaily.com/10625-businesses-collecting-data.html. Accessed 6 April 2022.

Friedman, A. (2007). *Sustainable residential Developlemts: Design principles for green Neighbourhoods*. McGraw-Hill.
Friedman, A. (2014). *Planning small and mid-sized towns; designing and retrofitting for sustainability*. Routledge.
Friedman, A. (2017). *Designing sustainable communities*. Bloomsbury.
Friedman, G. (2020). "Big-box retailers' profits surge as pandemic marches on". The New York Times. https://www.nytimes.com/2020/08/19/business/coronavirus-walmart-target-home-depot.html. Accessed April 26 2022.
Friedman, A. (2021). *Fundamentals of sustainable Urban Design*. Springer Publishing.
Friedman, A. (2022). *Designing innovative sustainable neighbourhoods*. Routledge.
Furszyfer Del Rioa, D. D., Sovacool, B. K., & Griffiths, S. (2021). Culture, energy and climate sustainability, and smart home technologies: A mixed methods comparison of four countries. *Energy and Climate Change, 2*, 2666–2787.
Gandini, A. (2015). The rise of coworking spaces: A literature review. *Ephemera, 15*(1), 193–205.
Georgijevic, A. (2016). "Opening up the library". The Globe and Mail. https://www.theglobeandmail.com/life/home-and-garden/design/the-reinvention-of-libraries-from-public-spaces-to-livingrooms/article29333752/. Accessed April 26 2022.
Gerace, A. (2013). "Virtual retirement communities see rising membership, Interest. *Senior Housing News*. https://seniorhousingnews.com/2013/07/22/virtual-retirement-communities-see-rising-membership-interest/. Accessed April 14 2022.
Gheorghiu, D., & Stefan, L. (2018). "Virtual museums: Dealing with cultural identity in the digital age". https://www.researchgate.net/publication/324151100_VIRTUAL_MUSEUMS_DEALING_WITH_CULTURAL_IDENTITY_IN_THE_DIGITAL_AGE. Accessed April 26 2022.
Gibson, E. (2019). "Sidewalk labs reveals masterplan for sidewalk Toronto Neighbourhood". Dezeen. https://www.dezeen.com/2019/06/28/toronto-tomorrow-sidewalk-labs-toronto/#. Accessed April 29 2022.
GISGeography. (2022). "Vector vs raster; What's the Difference Between GIS Spatial Data Types?" https://gisgeography.com/spatial-data-types-vector-raster/. Accessed April 26 2022.
Global Footprint Network. (n.d.-a). "Climate Change". https://www.footprintnetwork.org/our-work/climate-change/. Accessed 10 April 2022.
Global Footprint Network. (n.d.-b). "Data". https://data.footprintnetwork.org/#/. Accessed 10 April 2022.
Global Water Security. (n.d.). "Water security solutions Centre". https://globalwatersecurity.org/. Accessed 10 April 2022.
Goodwin, T. (2015). "The Battle is for the customer Interface". Tech Crunch. https://techcrunch.com/2015/03/03/in-the-age-of-disintermediation-the-battle-is-all-for-the-customer-interface/. Accessed April 26 2022.
Google Store. (n.d.). "Nest thermostat". https://store.google.com/ca/product/nest_thermostat?hl=en-GB. Accessed 6 April 2022.
Gottlieb, C., Grobovšek, J., Poschke, M., & Saltiel, F. (2021). Working from home in developing countries. *European Economic Review*, 103679.
Government of Canada. (2020). "Mental health support: Get help". https://www.canada.ca/en/public-health/services/mental-health-services/mental-health-get-help.html#:~:text=Call%201%2D800%2D668%2D,anonymous%20care%20from%20professional%20counsellors. Accessed April 14 2022.
Gow Hastings Architects. (2020). "Designing for higher education in a pandemic world". Canadian Architect. https://www.canadianarchitect.com/designing-for-higher-education-in-a-pandemic-world/. Accessed April 26 2022.
Gowling, W. L. G. (2019). "How will virtual reality change the way we plan a sustainable, smart city?" https://gowlingwlg.com/en/insights-resources/articles/2019/how-will-vr-change-the-way-we-plan-smart-cities/#:~:text=VR%20serves%20as%20a%20knight,on%20

infrastructure%20and%20the%20environment.&text=The%20larger%20augmented%20reality%2Fvirtual,billion%20in%20value%20by%202020. Accessed April 26 2022.
Grant, A. (n.d.). "Internet privacy by country". Best VP. https://bestvpn.org/privacy-index/. Accessed April 26 2022.
Green Design. (2022). "Sea train house". Northern Architecture. https://www.northernarchitecture.us/green-design-2/sea-train-house.html. Accessed April 14 2022.
Guidehouse Insights. (2018). "Smart Home Tech Moves to Low Income Households." https://guidehouseinsights.com/news-and-views/smart-home-tech-moves-to-low-income-households. Accessed April 26 2022.
Hardcastle, L., & Ogbogu, U. (2020). Virtual care: Enhancing access or harming care? In *Healthcare management Forum* (Vol. 33, No. 6, pp. 288–292). SAGE.
Harriet, T., Poku, K., & Emmanuel, A. K. (2013). An assessment of traffic congestion and its effect on productivity in Urban Ghana. *International Journal of Business and Social Science, 4*(3), 225–234. https://ijbssnet.com/journals/Vol_4_No_3_March_2013/25.pdf. Accessed April 10 2022.
HELEN. (n.d.). "Helsingin Energia's smart CHP/DH system – The most energy-efficient solution for heating Finland's capital". Helsingin Energia. https://www.districtenergyaward.org/wp-content/uploads/2012/10/Modernization_Finland_Helsinki_2011.pdf. Accessed April 14 2022.
Henkelman, J., & Sullivan, P. (2019). "Understanding landfill gas monitoring techniques". *Waste Today Magazine*. https://www.wastetodaymagazine.com/article/landfill-gas-monitoring-techniques/. Accessed April 14 2022.
I Mean It Creative. (2016). "Izmir City branding – Case study". https://www.youtube.com/watch?v=yCWTh8mQK3w&ab_channel=IMeanItCreative. Accessed April 26 2022.
IAIA Session Discussion. (2016). "Innovative communication of project and impacts". 36th Annual Conference of the International Association for Impact Assessment.
IEA. (n.d.). "Renewables". International Energy Agency IEA. https://www.iea.org/fuels-and-technologies/renewables. Accessed April 14 2022.
IESO. (n.d.). "A Smarter Grid". *Independent Electricity System Operator.* https://www.ieso.ca/en/Learn/Ontario-Power-System/A-Smarter-Grid. Accessed April 14 2022.
Instagram. (2020). "Keeping People Informed, Safe, and Supported on Instagram". from https://about.instagram.com/blog/announcements/coronavirus-keeping-people-safe-informed-and-supported-on-instagram. Accessed April 14 2022.
Internet Live Stats. (2022). "Internet Users". Internet Live Stats. https://www.internetlivestats.com/internet-users/. Accessed 6 April 2022.
Inventionland. (n.d.). "The history of Mobile apps". https://inventionland.com/inventing/the-history-of-mobile-apps/. Accessed 6 April 2022.
IPCC. (2018). "Summary for PolicyMakers". Special Report: Global Warming of 1.5 °C. Chap 00. https://www.ipcc.ch/sr15/chapter/spm/. Accessed 10 April 2022.
IPCC. (2021). "Special Report: Global Warming of 1.5 °C." https://www.ipcc.ch/sr15/. Accessed April 10 2022.
IPCC. (n.d.). "About the IPCC". https://www.ipcc.ch/about/. Accessed April 11 2022.
IRENA. (2015). "Renewable Energy integration in power grids". *International Renewable Energy Agency*. Abu Dhabi. https://www.irena.org/-/media/Files/IRENA/Agency/Publication/2015/IRENA-ETSAP_Tech_Brief_Power_Grid_Integration_2015.pdf. Accessed April 14 2022.
IRENA. (2021). "Renewable power generations costs in 2020". International Renewable Energy Agency, Abu Dhabi. https://www.irena.org/-/media/Files/IRENA/Agency/Publication/2021/Jun/IRENA_Power_Generation_Costs_2020.pdf?la=en&hash=755CB6E57667D53B629967EC7F9BE57A55D1CD78. Accessed April 14 2022.
Islam, N., & Want, R. (2014). Smartphones: Past, Present, and Future. *IEEE Pervasive Computing, 13*(4), 89–92. https://ieeexplore.ieee.org/abstract/document/6926722?casa_token=hfs_10n7E-xQAAAAA:nLpBni7XBVk9M4MEXn9aQ9u4r0wYm19QUgu0LGefqLsLpo8IAXCXWyMzwT_IT7DxsdChgqECng. Accessed 6 April 2022.

ITA. (2021). "What is Smart Home Irrigation?" https://integratedtechnologiesaustralia.com.au/resource-centre/what-is-smart-home-irrigation. Accessed April 26 2022.

Jacobs, J. (1961). *The death and life of great American cities*. Random House.

Jones, M. (2015). "The complete history of social media: A timeline of the invention of online networking". History Cooperative. https://historycooperative.org/the-history-of-social-media/. Accessed 6 April 2022.

Kahan, A. (2019). "EIA projects nearly 50% increase in world energy usage by 2050, led by growth in Asia". *U.S. Energy Information Administration EIA*. https://www.eia.gov/todayinenergy/detail.php?id=41433

Kaspersky. (2021). "How safe are smart homes?" AO Kaspersky Lab. https://www.kaspersky.com/resource-center/threats/how-safe-is-your-smart-home. Accessed April 26 2022.

Knox, P., & McCarthy, L. (2005). *Urbanization: An introduction to urban geography* (2nd ed.). Pearson Prentice Hall.

Kohls, L. R. (2001). *Survival kit for overseas living*. Intercultural Press.

Konyn, C. (2021). "What are carbon sinks?". Earth.Org. https://earth.org/carbon-sinks/. Accessed 10 April 2022.

Krishnamurthy, R. (2014). "Vertical farming: Singapore's solution to feed the local urban population". Permaculture Research Institute. https://www.permaculturenews.org/2014/07/25/vertical-farming-singapores-solution-feed-local-urban-population/. Accessed 6 April 2022.

Lam, P.-T. I., & Yang, W. (2020). Factors influencing the consideration of public-private partnerships (PPP) for smart city projects: Evidence from Hong Kong. *Cities, 99*, 102606. https://doi.org/10.1016/j.cities.2020.102606. Accessed 6 April 2022.

Land Transport Authority. (n.d.). "Land transport master plan 2040". https://www.lta.gov.sg/content/ltagov/en/who_we_are/our_work/land_transport_master_plan_2040.html. Accessed April 26 2022.

Lenz, B., Schreiber, J., & Stark, T. (2011). *"Optimising building operations" in sustainable building services*. DETAIL – Institut für internationale Architektur-Dokumentation GmbH & KG.

Li, J., Yang, X., & Sitzenfrei, R. (2020). Rethinking the framework of smart Water system: A review. *Water, 12*, 412.

Liceras, P. (2022). "A digital zebra crossing could be the future of our roads". Tomorrow. City. https://tomorrow.city/a/smart-zebra-crossing. Accessed April 26 2022.

Lichtenwoehrer, P., Erker, S., Zach, F., & Stoeglehner, G. (2019). Future compatibility of district heating in urban areas – A case study analysis in the context of integrated spatial and energy planning. *Energy, Sustainability and Society, 9*(1), 1–12.

Liszka, H. A., Steyer, T. E., & Hueston, W. J. (2006). Virtual medical care: How are our patients using online health information? *Journal of Community Health, 31*(5), 368–378.

Litman, T. (2017). "Introduction to multi-modal transportation planning: Principles and practices." Victoria Transport Policy Institute, Canada https://www.vtpi.org/multimodal_planning.pdf. Accessed April 10 2022.

Louis, J.-N., Caló, A., Leiviskä, K., & Pongrácz, E. (2015). Environmental impacts and benefits of smart home automation: Life cycle assessment of home Energy management system. *IFAC (International Federation of Automatic Control), 48*(1), 880–885.

Low, D. (2021). "Singapore is world's smartest city for the third year: IMD Smart City Index". The Straits Times. https://www.straitstimes.com/tech/tech-news/singapore-is-worlds-smartest-city-for-the-third-year-imd-smart-city-index#:~:text=SINGAPORE%20%2D%20Singapore%20is%20the%20smartest,this%20year's%20Smart%20City%20Index.&text=smart%22%20they%20are.-,A%20%22smart%20city%22%20is%20an%20urban%20setting%20that%20applies%20technology,of%20urbanisation%20for%20its%20citizens. Accessed 6 April 2022.

Mann, A. (2014). "What's up with that: Building bigger roads actually makes traffic worse." Wired. https://www.wired.com/2014/06/wuwt-traffic-induced-demand/. Accessed April 10 2022.

MapToasterTopo. (n.d.). "How GPS works". MapToaster. https://www.maptoaster.com/maptoaster-topo-nz/articles/how-gps-works/how-gps-works.html. Accessed 6 April 2022.

Martinez, O. (2017). "Keetwonen, Amsterdam's Cargo Container Student Housing Complex". *Containerpedia.* https://www.containerpedia.com/keetwonen-amsterdam-s-1000-cargo-container-student-housing-complex. Accessed April 14 2022.

Mayo Clinic. (2020). "Telehealth: Technology meets health care". Mayoclinic.org. https://www.mayoclinic.org/healthy-lifestyle/consumer-health/in-depth/telehealth/art-20044878. Accessed 6 April 2022.

McClellan, S., Jimenez, J. A., & Koutitas, G. (Eds.). (2018). *Smart Cities: Applications, technologies, standards, and driving factors* (p. 187). Springer.

MCI. (2022). "Digital Readiness". Ministry of Communication and Information. https://www.mci.gov.sg/en/portfolios/digital-readiness/get-digitally-ready. Accessed 6 April 2022.

Merkel, J. (2015). Coworking in the city. *Ephemera, 15*(2), 121–139.

Merriam-Webster. (n.d.). "Economy". Merriam-Webster.com Dictionary. https://www.merriam-webster.com/dictionary/economy. Accessed 10 April 2022.

Midgley, P. (2009). The role of smart Bike-sharing Systems in Urban Mobility. *Journeys, 2*(2). https://www.researchgate.net/publication/242235724_The_Role_of_Smart_Bike-sharing_Systems_in_Urban_Mobility Accessed April 10 2022.

Milton, K., Kelly, M. P., Baker, G., Cleland, C., Cope, A., Craig, N., Foster, C., Hunter, R., Kee, F., Kelly, P., Nightingale, G., Turner, K., Williams, A. J., Woodcock, J., & Jepson, R. (2021). Use of natural experimental studies to evaluate 20mph speed limits in two major UK cities. *Journal of Transport & Health, 22*. https://doi.org/10.1016/j.jth.2021.101141. Accessed April 26 2022.

Mirzabaev, A., Muñoz, P., & Zwick, S. (2020). The impact of urbanization on Austria's carbon footprint. *Journal of Cleaner Production, 263*. https://doi.org/10.1016/j.jclepro.2020.121326. Accessed 10 April 2022.

Mobi. (n.d.). "How it works". https://www.mobibikes.ca/en/how-it-works. Accessed April 25 2022.

Moore, O., & Younglai, R. (2021). "Toronto's PATH system faces crisis as companies announce permanent hybrid work arrangements". The Globe and Mail. https://www.theglobeandmail.com/business/article-torontos-path-system-faces-crisis-as-companies-announce-permanent/. Accessed April 26 2022.

Moreira, S. (2021). "What are garden cities?" Arch Daily. https://www.archdaily.com/961275/what-are-garden-cities. Accessed April 26 2022.

Moreno, J. (2017). The university is now on air: Broadcasting modern architecture[Exhibition]. Exhibited at the Canadian Center for Architecture November 15 2017–April 1 2018.

Moreno, C., Allam, Z., Chabaud, D., Gall, C., & Pratlong, F. (2021). Introducing the "15-Minute City": Sustainability, resilience and place identity in future post-pandemic cities. *Smart Cities, 4*(1), 93–111. https://doi.org/10.3390/smartcities4010006. Accessed April 26 2022.

Morgan, R. K. (2012). Environmental impact assessment: The state of the art. *Impact Assessment and Project Appraisal, 30*(1), 5–14.

NASA. (2014). "What is a satellite". https://www.nasa.gov/audience/forstudents/5-8/features/nasa-knows/what-is-a-satellite-58.html. Accessed 6 April 2022.

National Association of City Transportation Officials (NACTO). (2014). "Conventional Bike Lanes." https://nacto.org/publication/urban-bikeway-design-guide/bike-lanes/conventional-bike-lanes/#. Accessed April 2022.

National Centers for Environmental Information. (2021). "Global Climate Report – July 2021." https://www.ncdc.noaa.gov/sotc/global/202107. Accessed April 2022.

National Resources Canada. (2020). "Liquified Natural Gas". Government of Canada. https://www.nrcan.gc.ca/our-natural-resources/energy-sources-distribution/clean-fossil-fuels/natural-gas/liquefied-natural-gas/5679. Accessed 6 April 2022.

Naughton, J. (2016). The evolution of the internet: From military experiment to general purpose technology. *Journal of Cyber Policy, 1*(1). https://doi.org/10.1080/23738871.2016.1157619. Accessed 6 April 2022.

NEA. (n.d.). "Geothermal". National Energy Authority of Iceland". https://nea.is/geothermal/. Accessed April 14 2022.

NearSt. (n.d.). https://near.st/. Accessed April 26 2022.

Negash, S., & Meso, P. (n.d.). "Privacy concerns in the age of smart devices". Federal Trade Commission. https://www.ftc.gov/system/files/documents/public_comments/2017/11/00028-141795.pdf

Newman, D. (2020). "Distributed companies, work from home, and the technology enabling the change". Forbes. https://www.forbes.com/sites/danielnewman/2020/10/22/distributed-companies-work-from-home-and-the-technology-enabling-the-change/?sh=16b93c0033bf. Accessed April13 2022.

Nguyen, M. T., Nguyen, H. K., Vo-Lam, K. D., Nguyen, X. G., & Tran, M. T. (2016). Applying virtual reality in City planning. In S. Lackey & R. Shumaker (Eds.), *Virtual, augmented and mixed reality. VAMR 2016* (Lecture notes in Computer Science, Vol. 9740). Springer. https://doi.org/10.1007/978-3-319-39907-2_69. Accessed April 26 2022.

O'Neill, A. (2019). "Life expectancy in Canada, 1800–2020". *Statista*. https://www.statista.com/statistics/1041135/life-expectancy-canada-all-time/. Accessed April 14 2022.

Office of the Privacy Commissioner of Canada. (2020). "Smart devices and your privacy." Government of Canada. https://www.priv.gc.ca/en/privacy-topics/technology/02_05_d_72_iot/. Accessed April 26 2022.

Osborne Clark. (2016). "The Danish capital city will spend 60 million kroner (£6.2m) to implement intelligent transport systems (ITS), including smart traffic lights that prioritise buses and bicycles." https://smartcities.osborneclarke.com/its/copenhagen-to-install-smart-traffic-lights/. Accessed April 2022.

Palfrey, J. (2015). *BiblioTech: Why libraries matter more than ever in the age of Google*. Basic Books.

Pavid, K. (n.d.). "What is the Anthropocene and why does it matter?" Natural History Museum. https://www.nhm.ac.uk/discover/what-is-the-anthropocene.html. Accessed 10 April 2022.

Petriglieri, G., Ashford, S. J., & Wrzesniewski, A. (2018). "Thriving in the gig economy". Harvard Business Review. https://hbr.org/2018/03/thriving-in-the-gig-economy. Accessed April 13 2022.

Pfeffer, J. (2010). "Building sustainable organizations: The human factor". Stanford University Graduate School of Business". https://papers.ssrn.com/sol3/papers.cfm?abstract_id=1545977. Accessed 10 April 2022.

Phaneuf, A. (2021). "Latest trends in medical monitoring devices and wearable health technology". *Insider*. https://www.businessinsider.com/wearable-technology-healthcare-medical-devices. Accessed April 14 2022.

Pianezzi, D., Mori, Y., & Uddin, S. (2021). Public–private partnership in a smart city: A curious case in Japan. *International Review of Administrative Sciences*. https://doi.org/10.1177/00208523211051839. Accessed 6 April 2022.

Project Solar UK. (2018). "The history of renewably Energy: Where It all began". https://www.projectsolaruk.com/blog/history-renewable-energy-began/#:~:text=Renewable%20energy%20%E2%80%93%20defined%20as%20energy,mimic%20the%20workings%20behind%20hydropower. Accessed April 14 2022.

Proximity of Care. (2021). "Design Guide". Arup & the Bernard van Leer Foundation. https://www.proximityofcare.com/design. Accessed April 26 2022.

Pucher, J., & Buehler, R. (2008). Making cycling irresistible: Lessons from the Netherlands, Denmark and Germany. *Transport Reviews, 28*(4), 495–528. https://doi.org/10.1080/01441640701806612. Accessed April 10 2022.

Rahman, N. A. A., Sairi, I. H., Zizi, N. A. M., & Khalid, F. (2020). The Importance of Cybersecurity Education in School. *International Journal of Information and Education Technology, 10*. https://doi.org/10.18178/ijiet.2020.10.5.1393. Accessed April 26 2022.

Raw, L. (2009). "Striking a light: The Bryant and may matchwomen and their place in labour history". A&C Black.

Repair Café. (n.d.). https://www.repaircafe.org/en/about/. Accessed April 14 2022.

Rose, K. [@kacierose4]. (n.d.). [TikTok profile]. TikTok. https://www.tiktok.com/@kacierose4?fbclid=IwAR0bf61sDc9aF3D_2zpq8dCvWlbY-bE6A9sne1LHrI5FC0iTeuK-t33lve4. Accessed April 28 2022.

Rossow, A. (2018). "The birth of GDPR: What is it and what you need to know". Forbes. https://www.forbes.com/sites/andrewrossow/2018/05/25/the-birth-of-gdpr-what-is-it-and-what-you-need-to-know/?sh=5de41a4955e5. Accessed April 26 2022.

Sarkodie, S. A. (2021). Environmental performance, biocapacity, carbon & ecological footprint of nations: Drivers, trends and mitigation options. *Science of The Total Environment, 751*, 141912. https://doi.org/10.1016/j.scitotenv.2020.141912. Accessed 10 April 2022.

SAS. (n.d.). "Data Mining History & Current Advances". https://www.sas.com/en_us/insights/analytics/data-mining.html#:~:text=Data%20mining%20is%20the%20process,relationships%2C%20reduce%20risks%20and%20more. Accessed April 26 2022.

Satyro, W., Sacomano, J., Contador, J., & Telles, R. (2018). Planned obsolescence or planned resource depletion? A sustainable approach. *Journal of Cleaner Production, 195*, 744–752. https://doi.org/10.1016/j.jclepro.2018.05.222. Accessed April 14 2022.

Schaub, H. (2014). "Why senior citizens use the internet". https://www.brookings.edu/blog/techtank/2014/04/11/why-senior-citizens-use-the-internet/. Accessed April 14 2022.

Schraad-Tischler, D., & Seelkopf, L. (2015). "Concept and methodology – Sustainable governance indicators 2015". *Sustainable Governance Indicators.* https://www.sgi-network.org/docs/2020/basics/SGI_Concept.pdf. Accessed April 11 2022.

Schrotter, G., & Hürzeler, C. (2020). "The digital twin of the City of Zurich for urban planning". PFG. PG 99–112. https://doi.org/10.1007/s41064-020-00092-2. Accessed April 26 2022.

Schwab, K. (2016). "The fourth industrial revolution: What it means, how to respond". World Economic Forum. https://www.weforum.org/agenda/2016/01/the-fourth-industrial-revolution-what-it-means-and-how-to-respond/. Accessed 6 April 2022.

Second Harvest. (n.d.). Second Harvest food rescue. https://secondharvest.ca/foodrescue/app/terms-of-use/. Accessed 10 April 2022.

Sessums, Z. (2020). "We're really into the the Jetsons' space age style right now." Architectural Digest: Clever. https://www.architecturaldigest.com/story/the-jetsons-space-age-style. Accessed April 26 2022.

SGI. (n.d.). "About the SGI". Sustainable Governance Indicators. https://www.sgi-network.org/2020/About. Accessed 10 April 2022.

Shaheen, S. A., Cohen, A., & Martin, E. (2013). Public Bikesharing in North America. *Transportation Research Record Journal of the Transportation Research Board, 2387*(1), 83–92. https://doi.org/10.3141/2387-10. Accessed April 10 2022.

Shek, D. T., Chung, P. P., & Leung, H. (2015). Manufacturing economy vs. service economy: Implications for service leadership. *International Journal on Disability and Human Development, 14*(3), 205–215.

Sheriff, L. (2019). "This app enables you to make money off your own personal data". Forbes. https://www.forbes.com/sites/lucysherriff/2019/03/29/this-app-enables-you-to-make-money-off-your-own-personal-data/?sh=5d9f7aa429f6. Accessed 6 April 2022.

Shinbane, J. S., & Saxon, L. A. (2016). Digital monitoring and care: Virtual medicine. *Trends in Cardiovascular Medicine, 26*(8), 722–730.

Shively, B. (n.d.). "How much primary Energy is wasted before consumers see value from electricity?" Enerdynamics. https://www.enerdynamics.com/Energy-Currents_Blog/How-Much-Primary-Energy-Is-Wasted-Before-Consumers-See-Value-from-Electricity.aspx. Accessed April 14 2022.

Sim, A., Yaliraki, S. N., Barahona, M., & Stumpf, H. (2015). Great cities look small. *Journal of the Royal Society, Interface, 12*, 20150315. https://doi.org/10.1098/rsif.2015.0315. Accessed April 26 2022.

Sinclair, A. J., Peirson-Smith, T. J., & Boerchers, M. (2017). Environmental assessments in the internet age: The role of e-governance and social media in creating platforms for meaningful participation. *Impact Assessment and Project Appraisal, 35*(2), 148–157.

Sitra. (2011). "World-class sustainable solutions from Finland". *Finnish Innovation Fund Sitra, VTT Technical Research Centre of Finland, Tekes.* https://www.sitra.fi/en/publications/world-class-sustainable-solutions-finland/. Accessed 10 April 2022.

Sitra. (2020). "Connecting food businesses and charities to capture surplus food". The Finnish Innovation Fund Sitra. https://www.sitra.fi/en/cases/connecting-food-businesses-and-charities-to-capture-surplus-food/. Accessed 10 April 2022.

Smart Nation Singapore. (n.d.-a). "Pillars of a Smart Nation". https://www.smartnation.gov.sg/about-smart-nation/pillars-of-smart-nation. Accessed 6 April 2022.

Smart Nation Singapore. (n.d.-b). "Achievements". https://www.smartnation.gov.sg/about-smart-nation/our-journey/achievements. Accessed 6 April 2022.

SmartGrid. (n.d.). "The Smart Grid". U.S. Department of Energy. https://www.smartgrid.gov/the_smart_grid/smart_grid.html. Accessed April 14 2022.

Smith, B. (2021). "How TikTok reads your mind". New York Times. https://www.nytimes.com/2021/12/05/business/media/tiktok-algorithm.html. Accessed April 28 2022.

Smithsonian. (n.d.). "City and suburb". National Museum of American History. https://americanhistory.si.edu/america-on-the-move/city-and-suburb. Accessed April 26 2022.

Soares Machado, C. A., Marie de Salles Hue, N. P., Berssaneti, F. T., & Quintanilha, J. A. (2018). An overview of shared mobility. *Sustainability, 10*(12). https://doi.org/10.3390/su10124342. Accessed April 10 2022.

Sobolevsky, S., Levitskaya, E., Chan, E., Postel, M., & Kontokosta, C. (2018). *Impact of Bike sharing in New York City*. Submitted to Cornell University.

Société de Transport de Montréal. (n.d.). "iBUS – Real time." https://www.stm.info/en/about/major_projects/completed-major-projects/ibus-real-time. Accessed April 10 2022.

Sociotio. (2020). "The Facts About Ageing in Place." https://www.scioto.com/. Accessed April 26 2022.

Statistics Canada. (2017). "Journey to work: Key results from the 2016 Census." https://www150.statcan.gc.ca/n1/daily-quotidien/171129/dq171129c-eng.htm. Accessed April 13 2022.

Statistics Canada. (2021). "Working From Home in Canada: What have we learned so far?" https://www150.statcan.gc.ca/n1/pub/36-28-0001/2021010/article/00001-eng.htm. Accessed April 13 2022.

Steenson, R. (2019). "The history of Geographic Information Systems (GIS)". The Chartered Institute for IT. https://www.bcs.org/articles-opinion-and-research/the-history-of-geographic-information-systems-gis/. Accessed April 26 2022.

Stobierskie, T. (2021). "4 impactful sustainable business practices that can makes a difference". Harvard Business School Online. https://online.hbs.edu/blog/post/sustainable-business-practices. Accessed 10 April 2022.

Student Energy. (n.d.). "Electrical Grid". https://studentenergy.org/distribution/electrical-grid/

Sustainia. (2018). "Copenhagen: Smart traffic signals boost cycling." Global Opportunity Explorer. https://goexplorer.org/copenhagen-smart-traffic-signals-boost-cycling/. Accessed April 10 2022.

Taber, J. (2014). "Halifax looks forward to the opening of its very ow library of the future". The Globe and Mail. https://www.theglobeandmail.com/news/national/halifax-looks-forward-to-the-opening-of-its-very-own-library-of-the-future/article20090514/. Accessed April 26 2022.

Tattrie, J. (2019). "Canada and the digital economy". The Canadian Encyclopedia. https://www.thecanadianencyclopedia.ca/en/article/digital-economy-in-canada. Accessed April 26 2022.

Techopedia. (n.d.). "Smart device". Techopedia Dictionary. https://www.techopedia.com/definition/31463/smart-device. Accessed 6 April 2022.

Temming, M. (2021). "New recycling technologies could keep more plastic out of landfills". Science News for Students. https://www.sciencenewsforstudents.org/article/new-recycling-technologies-could-keep-more-plastic-out-of-landfills. Accessed April 14 2022.

Thakuriah, P., Tilahun, N. Y., & Zellner, M. (2017). Big data and urban informatics: Innovations and challenges to urban planning and knowledge discovery. In *Seeing cities through big data.*

Springer Geography. Springer. https://doi.org/10.1007/978-3-319-40902-3_2. Accessed April 26 2022.

Toronto Public Library [TPI]. (n.d.). "Live & Online Programs". https://www.torontopubliclibrary.ca/programs-and-classes/. Accessed April 27 2022.

Townsend, A. M. (2014). *Smart cities: Big data, civic hackers, and the quest for a new utopia*. W.W. Norton & Company.

Transport Policy Advisory Services on behalf of the Federal Ministry for Economic Cooperation and Development. (2009). Cycling-inclusive policy development: A handbook. Utrecht. https://sutp.org/publications/cycling-inclusive-policy-development-a-handbook/. Accessed April 10 2022.

Tuohy, J. P. (2021). "The best smart sprinkler controller". NY Times. https://www.nytimes.com/wirecutter/reviews/best-smart-sprinkler-controller/. Accessed April 26 2022.

Turner-Lee, N. (2020a). "Bridging digital divides between schools and communities". Brookings. https://www.brookings.edu/research/bridging-digital-divides-between-schools-and-communities/. Accessed April 26 2022.

Turner-Lee, N. (2020b). "What the coronavirus reveals about the digital divide between schools and communities". Brookings. https://www.brookings.edu/blog/techtank/2020/03/17/what-the-coronavirus-reveals-about-the-digital-divide-between-schools-and-communities/. Accessed April 26 2022.

TWI. (n.d.). "What is a smart city? – Definition and examples". TWI Global – The Welding Institute. https://www.twi-global.com/technical-knowledge/faqs/what-is-a-smart-city#:~:text=For%20example%2C%20smart%20parking%20can,by%20a%20smart%20city%20infrastructure. Accessed 6 April 2022.

U.S. Department of Transportation (DOT). (2010). Transportation's Role in Reducing U.S. Greenhouse Gas Emissions: Volume 1. Report to U.S. Congress. https://rosap.ntl.bts.gov/view/dot/17789. Accessed April 10 2022.

U.S. Department of Transportation Federal Highway Administration. (2016). Shared Mobility: Current Practices and Guiding Principles. Washington, DC. https://ops.fhwa.dot.gov/publications/fhwahop16022/index.htm. Accessed April 10 2022.

U.S. Energy Information Administration. (2011). "Residential Energy Consumption Survey (RECS)." https://www.eia.gov/consumption/residential/reports/2009/air-conditioning.php. Accessed April 26 2022.

UCSUSA. (2022). "UCS satellite database". Union of Concerned Scientists. https://www.ucsusa.org/resources/satellite-database. Accessed 6 April 2022.

UN. (1972). "Report of the United nations Conference on the Human Environment". *United* Nations. https://www.un.org/ga/search/view_doc.asp?symbol=A/CONF.48/14/REV.1. Accessed 10 April 2022.

UN (2021). "Digital Economy Report 2021". United Nations Conference on Trade and Development. https://unctad.org/webflyer/digital-economy-report-2021. Accessed 10 April 2022.

UN. (n.d.-a). "11: Sustainable Cities and Communities". https://www.un.org/sustainabledevelopment/cities/. Accessed April 10 2022.

UN. (n.d.-a). "11: Sustainable cities and communities". https://www.un.org/sustainabledevelopment/cities/. Accessed April 15 2022.

UN. (n.d.-b). "Sustainability". https://www.un.org/en/academic-impact/sustainability#:~:text=In%201987%2C%20the%20United%20Nations,development%20needs%2C%20but%20with%20the. Accessed April 10 2022.

UN. (n.d.-b). "11: Sustainable cities and communities". https://www.un.org/sustainabledevelopment/cities/. Accessed April 2022.

UN. (n.d.-c). "The 17 Goals". *United Nations Department of Economic and Social Affairs – Sustainable Development.* https://unric.org/en/united-nations-sustainable-development-goals/#:~:text=The%20SDGs%20are%20integrated%E2%80%94that,for%20those%20furthest%20behind%20first. Accessed April 10 2022.

UN. (n.d.-c). "Digital Economy Report 2021". United Nations Conference on Trade and Development. https://unctad.org/webflyer/digital-economy-report-2021. Accessed 10 April 2022.

UN. (n.d.-d). "What is Climate Change?" https://www.un.org/en/climatechange/what-is-climate-change. Accessed April 10 2022.

UN. (n.d.-d). "Goal 11: Make cities inclusive, safe, resilient and sustainable". United Nations. https://www.un.org/sustainabledevelopment/cities/. Accessed April 26 2022.

UN Water. (2021). "Water scarcity." United Nations. https://www.unwater.org/water-facts/scarcity/. Accessed April 26 2022.

UNECE. (n.d.). "Sustainable smart cities." UNECE. https://unece.org/housing/sustainable-smart-cities#:~:text=A%20smart%20sustainable%20city%20is,as%20well%20as%20cultural%20aspects. Accessed 6 April 2022.

UNFCCC. (n.d.). "The Paris agreement". *United Nations Climate Change UNFCCC*. https://unfccc.int/process-and-meetings/the-paris-agreement/the-paris-agreement. Accessed April 14 2022.

United Nations. (n.d.). "The Paris agreement." UNFCCC Sites and platforms. https://unfccc.int/process-and-meetings/the-paris-agreement/the-paris-agreement. Accessed April 10 2022.

United States. (2021a). "Electricity explained". *U.S. Energy Information Administration EIA*. https://www.eia.gov/energyexplained/electricity/delivery-to-consumers.php. Accessed April 14 2022.

United States. (2021b). "Renewable energy explained". U.S. Energy Information Administration EIA. https://www.eia.gov/energyexplained/renewable-sources/

United States. (n.d.). "Benefits of Residential Solar Energy". *U.S. Department of Energy*. https://www.energy.gov/energysaver/benefits-residential-solar-electricity. Accessed April 14 2022.

United States Environmental Protection Agency (EPA). (2019). "Greenhouse Gas Emissions – Sources of Greenhouse Gas Emissions." https://www.epa.gov/ghgemissions/sources-greenhouse-gas-emissions. Accessed April 10 2022.

University of Gävle. (2018). "Economic Sustainability". https://www.hig.se/Ext/En/University-of-Gavle/About-the-University/Environmental-Work/What-is-sustainable-development-at-HiG/Economic-sustainability.html. Accessed April 10 2022.

University of Toronto. (2017). "Resilience planning guide". *John H. Daniels Faculty of Architecture. Landscape and Design*.

Van Dessel, A. (2019). "The use of virtual reality in urban planning". University of Antwerp, Faculty of Design Sciences. https://scriptiebank.be/sites/default/files/thesis/2019-09/VanDessel-Anke_Masterproef%20%28incl%20kaft%29.pdf. Accessed April 26 2022.

Varrella, S. (2021). "Canada – resident population, by gender and age group 2020". *Statista*. https://www.statista.com/statistics/444858/canada-resident-population-by-gender-and-age-group/#:~:text=The%20average%20age%20of%20the,males%20and%20females%2C%20are%20single. Accessed April 14 2022.

Victoria State Government. (n.d.). "20-minute neighbourhoods". https://www.planning.vic.gov.au/policy-and-strategy/planning-for-melbourne/plan-melbourne/20-minute-neighbourhoods. Accessed April 26 2022.

Vivint Solar. (n.d.). "What is net metering?" https://www.vivintsolar.com/learning-center/what-is-net-metering. Accessed April 14 2022.

Wagner, K. (2018). "City noise might be making you sick." The Atlantic. www.theatlantic.com/technology/archive/2018/02/city-noise-might-be-making-you-sick/553385/. Accessed April 10 2022.

Water Science School. (2019). "How Much Water is There on Earth?" *U.S. Geological Survey*. https://www.usgs.gov/special-topics/water-science-school/science/how-much-water-there-earth#:~:text=a%20watery%20place.-,But%20just%20how%20much%20water%20exists%20on%2C%20in%2C%20and%20above,percent%20of%20all%20Earth's%20water. Accessed April 10 2022.

Wheeler, S. (2004). "Planning for sustainability: Creating livable equitable and ecological communities". https://books.google.ca/books/about/Planning_for_Sustainability.html?id=rtghgHKV6pwC&source=kp_book_description&redir_esc=y. Accessed April 10 2022.

WHO. (n.d.). "Air pollution". World Health Organization. https://www.who.int/health-topics/air-pollution#tab=tab_2. Accessed April 10 2022.

Wikipedia. (2022a). "Encyclopædia Britannica". https://en.wikipedia.org/wiki/Encyclop%C3%A6dia_Britannica. Accessed April 27 2022.

Wikipedia. (2022b). "Wikipedia". https://en.wikipedia.org/wiki/Wikipedia#Language_editions. Accessed April 27 2022.

Wilczek, M. (2022). "Poland's "spatial chaos" caused by poor planning wastes 84bn zloty per year: Report". Notes from Poland". https://notesfrompoland.com/2022/01/20/polands-spatial-chaos-caused-by-poor-planning-wastes-84bn-zloty-per-year-report/?fbclid=IwAR0u6I6pU9fZGoHrgflrcl95sz81B5cxsz3sqBdqCvWGpFU5DnEQqptotDU. Accessed April 26 2022.

Wood, E. (2020). "What is a microgrid?" Microgrid Knowledge. https://microgridknowledge.com/microgrid-defined/#:~:text=A%20microgrid%20is%20a%20self,generators. Accessed April 14 2022.

Woon, J. (2018). "Technology and governance in Singapore's smart nation initiative". ASH Center for Democratic Governance and Innovation. Harvard Kennedy School. https://ash.harvard.edu/publications/technology-and-governance-singapore%E2%80%99s-smart-nation-initiative. Accessed April 10 2022.

World Bank. (n.d.). "What a Waste 2.0". The World Bank. https://datatopics.worldbank.org/what-a-waste/trends_in_solid_waste_management.html#:~:text=The%20world%20generates%202.01%20billion,from%200.11%20to%204.54%20kilograms. Accessed April 14 2022.

World Bank Group. (2020). "Individuals using the Internet (% of population)". https://data.worldbank.org/indicator/IT.NET.USER.ZS. Accessed April 13 2022.

World Economic Forum. (2015). "Top ten urban innovations".https://www.weforum.org/agenda/2017/02/cities-must-tirelessly-innovative-to-respond-to-their-challenges/. Accessed April 26 2022.

World Health Organization. (2017). "Cardiovascular diseases (CVDs)". https://www.who.int/news-room/fact-sheets/detail/cardiovascular-diseases-(cvds). Accessed April 14 2022.

World Health Organization: Regional Office for Europe. (2011). Burden of disease from environmental noise: Quantification of healthy life years lost in Europe. JRC European Commission. https://apps.who.int/iris/handle/10665/326424. Accessed April 10 2022.

World Wildlife Foundation. (n.d.). "Ecological Footprint". https://wwf.panda.org/discover/knowledge_hub/teacher_resources/webfieldtrips/ecological_balance/eco_footprint/. Accessed April 10 2022.

Zhang, J. (2016). The dark side of virtual office and job satisfaction. *International Journal of Business and Management, 11*(2), 40.

Index

A. Friedman, *The Sustainable Digital City*,
https://doi.org/10.1007/978-3-031-25488-8

Springer Theses

Recognizing Outstanding Ph.D. Research

Aims and Scope

The series "Springer Theses" brings together a selection of the very best Ph.D. theses from around the world and across the physical sciences. Nominated and endorsed by two recognized specialists, each published volume has been selected for its scientific excellence and the high impact of its contents for the pertinent field of research. For greater accessibility to non-specialists, the published versions include an extended introduction, as well as a foreword by the student's supervisor explaining the special relevance of the work for the field. As a whole, the series will provide a valuable resource both for newcomers to the research fields described, and for other scientists seeking detailed background information on special questions. Finally, it provides an accredited documentation of the valuable contributions made by today's younger generation of scientists.

Theses may be nominated for publication in this series by heads of department at internationally leading universities or institutes and should fulfill all of the following criteria

- They must be written in good English.
- The topic should fall within the confines of Chemistry, Physics, Earth Sciences, Engineering and related interdisciplinary fields such as Materials, Nanoscience, Chemical Engineering, Complex Systems and Biophysics.
- The work reported in the thesis must represent a significant scientific advance.
- If the thesis includes previously published material, permission to reproduce this must be gained from the respective copyright holder (a maximum 30% of the thesis should be a verbatim reproduction from the author's previous publications).
- They must have been examined and passed during the 12 months prior to nomination.
- Each thesis should include a foreword by the supervisor outlining the significance of its content.
- The theses should have a clearly defined structure including an introduction accessible to new PhD students and scientists not expert in the relevant field.

Indexed by zbMATH.